Coastal Area Management and Development

Related Pergamon Titles of Interest

Books

COLLINS *et al.*
Industrialised Embayments and their Environmental Problems
(A Case Study of Swansea Bay)

DAVIDSON
Planning and the Rural Environment

FROST
The Impact of Regional Policy

GILDER
Rival Planning Policies: An Economic Appraisal

HOLY
Erosion and Environment

HOYLE
Cityport Industrialization and Regional Development

WHO/UNEP
Waste Discharge into the Marine Environment
(Principles and Guidelines for the Mediterranean Action Plan)

YEN & WALSH
Energy and Resource Development of Continental Margins

Journals

Continental Shelf Research

Environment International

Marine Pollution Bulletin

Water Research

Water Science & Technology

Water Supply & Management

Coastal Area Management and Development

UNITED NATIONS
DEPARTMENT OF INTERNATIONAL ECONOMIC AND SOCIAL AFFAIRS

Ocean Economics and Technology Branch

PERGAMON PRESS

OXFORD · NEW YORK · TORONTO · SYDNEY · PARIS · FRANKFURT

U.K.	Pergamon Press Ltd., Headington Hill Hall, Oxford OX3 0BW, England
U.S.A.	Pergamon Press Inc., Maxwell House, Fairview Park, Elmsford, New York 10523, U.S.A.
CANADA	Pergamon Press Canada Ltd., Suite 104, 150 Consumers Rd., Willowdale, Ontario M2J 1P9, Canada
AUSTRALIA	Pergamon Press (Aust.) Pty. Ltd., P.O. Box 544, Potts Point, N.S.W. 2011, Australia
FRANCE	Pergamon Press SARL, 24 rue des Ecoles, 75240 Paris, Cedex 05, France
FEDERAL REPUBLIC OF GERMANY	Pergamon Press GmbH, 6242 Kronberg-Taunus, Hammerweg 6, Federal Republic of Germany

First edition 1982

Library of Congress Cataloging in Publication Data

Main entry under title:
Coastal area management and development.
Includes bibliographical references.
1. Coastal zone management. 2. Economic development.
3. Underdeveloped areas. 4. Coastal zone management—
Law and legislation. I. United Nations. Ocean
Economics and Technology Branch.
HT391.C497 1982 333.91'715 82-14991

British Library Cataloguing in Publication Data

United Nations. *Ocean Economics and Technology Branch*
Coastal area management and development.
1. Coastal zone management
2. Regional planning
I. Title
361.6'1 HT391

ISBN 0-08-023393-7

In order to make this volume available as economically and as rapidly as possible the typescript has been reproduced in its original form. This method unfortunately has its typographical limitations but it is hoped that they in no way distract the reader.

Printed in Great Britain by A. Wheaton & Co. Ltd., Exeter

Preface

In 1973, the Economic and Social Council of the United Nations described coastal areas as a "valuable national possession" and declared that their "proper management and development" constitutes an important factor in national development planning. This tenet has provided the basis and impetus for the implementation of a United Nations programme of activities in coastal area management and development. The present publication, prepared in two parts by the Ocean Economics and Technology Branch (OETB) of the United Nations Department of International Economic and Social Affairs, attempts to give operational significance to the concept of proper management by spanning the gap between theory and application.

Part I entitled "Establishing a National Programme: Economic and Technical Matters", sets out to provide guidance to planners and decision-makers particularly in developing countries on how an effective, long-term programme of coastal area management can be implemented within the overall framework of development planning. In order to provide researchers, planners, and policy-makers with a guide to international literature on economic and technical aspects of coastal area management and development, a number of references are included, selected on the basis of their quality of work, timeliness and accessibility to a wide audience.

Part I does not set out to provide a treatise on development planning methods, although "coastal area management" is certainly a part of national development planning. Nor does it discuss in detail any of the coastal activities and uses which together comprise 'coastal area development'. These activities, some of which are dealt with in Part II, are, in general, treated by one or more organizations within the United Nations system; for example, FAO's programme in agriculture, be it inland or coastal, is sufficiently detailed and comprehensive to make it unnecessary to treat coastal agriculture *per se*. Similar considerations hold for the other categories of activities.

However, in preparing Part I, a thorough survey was made of the voluminous literature on the theory of planning and management and its applicability in the coastal context, including case studies. Account has also been taken of the 'sectoral' work of United Nations Organizations and of the work of the large number of intergovernmental bodies, international professional societies, non-governmental organizations and scientific expert groups that are addressing themselves to various aspects of coastal area management and development. An earlier study entitled "Coastal Area Management and Development" prepared by the Ocean Economics and Technology Branch for the Economic and Social Council in 1975 (document E/5648, 8 May 1975) provided the conceptual framework for the present undertaking.

A review of all these going efforts reveals varying objectives, differing emphases and a multitude of approaches to the analysis of the problems. However, in Part I, coastal management and development is discussed in a restricted and well-defined sense and the approach taken is an economic one. No existing publication has to date covered the same ground with the same viewpoint, nor has much of the literature been aimed at practitioners in developing countries.

This approach is not intended to deny the importance of social, political and organizational factors. However, a number of ongoing efforts in the United Nations system treat these factors more comprehensively than could be done in a single volume. The United Nations activities in rural development, a unified approach to social and economic development planning, and the UNEP studies on eco-development are all relevant to the broad problems.

Clearly, decisions are never made solely on an economic basis, for that would deny the existence of legitimiate social goals, such as a reduction of inequality among individuals or groups. However, every society -- capitalist, socialist, or mixed -- has a limited pool of resources with which it can achieve a mixture of economic growth, social development, and environmental conservation. Hence, whatever a country's objectives may be with respect to the exploitation of the coastal area and its resources, economic criteria will always be a necessary part of policy and decision-making, although not necessarily the sole basis for action.

The first draft of Part I of this publication was sent for review and comments to over a dozen experts around the world as well as to several organizations within the United Nations system including the Food and Agriculture Organization (FAO), the United Nations Educational, Scientific and Cultural Organization (UNESCO), the Inter-Governmental Maritime Consultative Organization (IMCO), the International Atomic Energy Agency (IAEA), the World Meterological Organization (WMO), the World Health Organization (WHO), the International Telecommunication Union (ITU), Economic Commission for Africa (ECA), Economic and Social Commission for Asia and the Pacific (ESCAP), Economic Commission for Latin America (ECLA), Economic Commission for West Asia (ECWA), Economic Commission for Europe (ECE) and United Nations Environment Programme (UNEP). The final version reflects their assistance, which is gratefully acknowledged. Particular thanks are due to Professor Rhodes Fairbridge, Professor Saul Saila and Dr. H.A. Cole for their technical input and advice.

Complementing the emphasis in Part I on the economic and technical dimensions of establishing a national programme in coastal area management and development, Part II of the present publication focusses on the legislative and regulatory aspects. It endeavours to provide a range of examples of the kind of legislative and policy measures and approaches that have been taken or may be taken in different countries to deal with some specific issues. It consists of three sections.

The first section contains examples of critical issues and planning aspects in twenty-five countries. The information is largely based on published sources and communications received from the respective Governments. The second section contains a survey of legislative measures in forty nations, promulgated either specifically for the purpose of coastal area management, or for general purposes that are none the less relevant to certain important coastal activities. For the sake of convenience these measures are grouped under five chapters: adaptation of town and country planning techniques to the coastal areas; measures for protection of shorelines; natural resource management approach; special measures for critical sections; and establishment of national guidelines for management programmes. Information used in this section is based on communications received from the respective governments and from materials supplied by consultants for the following countries: Kenya and Tanzania (Mr. G. Munoru); India, Pakistan and Sri Lanka (Mr. J.D.M. Derrett); Argentina, Brazil, Ecuador, Mexico, Peru and Venezuela (Ms. B.M. Carl); Bulgaria, German Democratic Republic, Poland, Romania and the Soviet Union (Messrs. R. Bierzanek and I. Bulska, and Ms. B. Czchowska). Their valuable contributions are hereby acknowledged. Chapter 15 is a summary of the United Nations' activities and programmes in coastal area management and development.

On the basis of an analysis of the survey results, a number of specific needs are identified in the third section. Many government agencies seem concerned mainly with fostering industrial or economic development and they may therefore be expected to show greater concern with developmental goals than with other competing factors or values. It was found useful to include some general legislative and policy guidelines for managing the coastal area. They are set out in Chapter 16. These guidelines stress that development should not disregard environmental considerations and that both factors should be considered together to attain sound resource management.

Three specific issues seem to deserve special attention in the context of coastal management: the establishment of criteria for selecting appropriate locations for industries, the control of oil pollution in the coastal area, and the establishment of criteria for construction setback lines. Experience shows that proper siting of industrial facilities can often ensure better development, avoid costly alteration and reduce adverse effects on other attributes (e.g. scenic, water quality or conservation of wetland). So far, not enough attention seems to have been given to the interactions which may exist. Chapter 17 therefore deals with the question of selection of sites.

Oil pollution, the second issue, may constitute a major threat to coastal amenities and utilization of other resources, and must be effectively controlled. Some basic guidelines for the purpose of regulation are thus provided in Chapter 18. Many countries are facing the difficult problem of erosion. In coastal areas, this natural process is often accelerated by man-made structures, such as breakwaters, jetties, etc. Engineering works are not only extremely costly but may not offer a permanent solution. The establishment of a coastal construction setback line is therefore considered a valuable means to cope with the problem. The setback line is not a primary corrective measure for erosion, but it serves two main purposes: (i) to project future development by limiting it outside the probable erosive zone deformed by the setback lines and (ii) to halt additional erosion that can be triggered by new construction. Chapter 19 provides certain guidelines regarding the establishment of a construction setback line.

In conclusion, the present publication, the first of its kind prepared by the United Nations, does not attempt to provide a comprehensive or definitive treatment. Rather, it seeks to provide certain insights into the requirements, problems and opportunities associated with coastal area management and development and practical guidance for developing national programmes.

Contents

PART I

Establishing a National Programme: Economic and Technical Matters

CHAPTER 1

Part I: Introduction

The coastal area is the area of interaction between the land and the sea. It therefore includes terrestrial and marine resources, both renewable and non-renewable. For non-renewable resources, such as mineral deposits, the economic problem which a society must face is how to mine the deposit at least cost and over what period of time to deplete it. In the case of a renewable resource, such as a fish stock, the economic problem is how to harvest the stock and to choose a rate which allows the stock to maintain its productivity. In addition to these resource allocation and utilization problems, environmental interactions are an important factor in the coastal area, and negative environmental effects of development in the coastal area are not considered in private economic assessments, but do impose a real cost on society.

The basic objective of coastal area management is to ensure a level and type of development of an area and its resources which is consistent with the continuing productivity and viability of the natural systems upon which the productivity of the area is based. In current terminology, coastal management is "integrated resource management" in a specific physical context, and it is also a form of what is called "environmental management".

The purpose of this publication is to discuss specific economic, scientific and technical problems associated with establishing a national programme of coastal management, particularly in developing countries. Guidelines are given for various components of a national programme. No claims are made to universality of coverage in terms of topics or in terms of problems likely to be encountered in establishing and managing a programme.

Coastal management problems can be grouped into three categories:

(1) problems arising from <u>interactions</u> between different development activities;

(2) problems arising from the <u>modification of coastal processes or the destruction of critical environmental units</u>;

(3) problems arising from the <u>pollution of coastal and estuarine waters</u>.

Interactions between activities may occur because of the proximity of two activities in the limited coastal space. In the course of dredging a harbour channel, for example, submarine pipelines or cables may be dug up or cut; this type of interaction occurs because of the failure of engineering or planning processes within one or both activities. Attention will not be given to these sectoral problems.

The second and third categories of coastal management problems referred to above, cut across the interests of any single economic activity or sector and affect the national economy in a wide sense, i.e. they affect human well-being and the quality of life. Also, these problems can affect the well-being of people for many years beyond the time considered when making economic evaluations.

In developing countries there is growing pressure to develop seemingly marginal lands to meet the needs of increasing populations and to improve living standards. Often these marginal lands are in the coastal area, and their transformation can have serious short and long-term implications. The case of the coastal swamplands in Sumatra /36/ is a very instructive example. A proposed scheme for irrigated agriculture with attendant use of chemical fertilizers was, under the usual economic and technical analysis, shown to be beneficial. However, it raised very real questions about the fate of rich fisheries which would be affected by the agricultural run-offs and about the alternative uses of the mangrove forests, which were scheduled for clearing under the proposal. There were serious and irreversible long-term implications following from the agricultural development proposal. An integrated resource management approach, i.e. a coastal management framework, proved to be a useful way of handling important socio-economic and environmental decisions which would not have otherwise received adequate consideration under sectoral decision-making processes.

Pollution of coastal and estuarine waters can pose hazards to human health, lead to damage or destruction of fisheries, and reduce the amenity value of natural areas for recreation and conservation. Coastal waters have a capacity to receive wastes from human activities and to absorb and process them harmlessly; this capacity is a natural resource, and it would not be rational to leave it unused. However, beyond some threshold level which varies with physical characteristics of the particular coastal area and with the waste substance, introduction of additional waste results in 'pollution'. Reduction in the amount of coastal pollution will always imply a resource cost, either in terms of forgone production or in terms of resources for environmental control and monitoring. Increases in environmental quality (reduction of pollution) are obtained by 'trading off' resources which could have been used elsewhere. Within the coastal management process, these types of decisions will have to be made, either for the whole coastal area or for certain critical areas.

The types of problem encountered in coastal management are encountered throughout the economy, and they are not unique to the coastal area. However, coastal area management assumes its importance because of the unusual productivity of the area and its ecosystems and because of man's extensive and growing use of these areas for settlement, agriculture, trade, industry, and amenity. An earlier United Nations study 1/ stresses these points in more detail.

It is the purpose of this part of the publication to give guidelines for the basic processes of coastal management which should be incorporated into a national programme. Considerations of costs and amounts of required resources cannot be made explicit since, among other things, they depend on the amount of existing technical and managerial infrastructure in place, and on the degree of importance

---

1/ "Coastal Area Management and Development: Report of the Secretary-General", E/5648, 1974.

attached to coastal problems in a particular country. It is intended that interested readers can pursue detailed professional interests through the selected references given at the end of this Part.

CHAPTER 2

Economic Principles

In both developed and developing countries, there is intense pressure to develop the coastal areas for tourism, mineral extraction, fishing, agriculture, ports, and industrial development. Yet the need to ensure the continuing productivity of coastal ecosystems creates pressure to restrict development for the sake of the environment. So the coastal area becomes a specific battleground for the debate between 'developers' and 'environmentalists'. This debate often takes the form of absolute conflict, i.e. one is either for development (environment) or one is against it. An economist would say that this debate is not about conflict but about choice and the allocation of scarce resources to a set of goals. Economic principles are very relevant to the coastal area, but they are often forgotten in this context.

At any moment in time, a society possesses a fixed endowment of productive factors, i.e. capital, labour and natural resources. For simplicity we assume society produces only three categories of 'products', viz. food, clothing, and environmental quality, for which we assume there is some index measuring the 'quality of life'. 1/ Under a set of reasonable assumptions about the relationship between factor inputs and outputs, it is possible to derive the set of all possible combinations of the three products that society can technically produce. This set of possibilities is described in this example by a three-dimensional surface, as shown in Fig. 1.

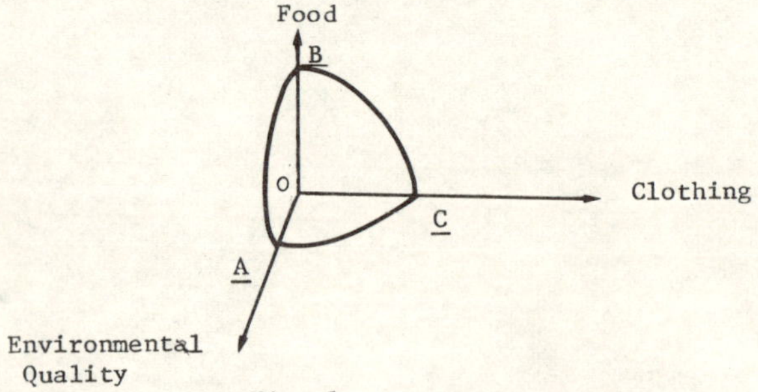

Fig. 1

1/ See /40/ and /86/ for details on environmental indices.

For technical reasons, this surface will be convex to the origin, i.e. it is bowed outward.2/ All interior points are wasteful of society's resources because, by simply reallocating its factors of production, society can have more of all three products without sacrificing any resources. Points outside this frontier are not attainable, given the initial endowment of society.

The actual point at which society produces, and therefore the total output of each product, depends on the preferences of its citizens. Thus for example, two societies may have the same technical production possibilities (the three-dimensional surface), but one society might value food highly while the other might value environmental quality. An economist would not be able to say unequivocably that one society's preferences were better than the other's.

Using the diagram, it is possible to show that a policy directive such as "maximizing economic efficiency and environmental quality" is not meaningful. Environmental quality is maximized at point A, where society produces nothing, pollutes not at all, and puts all its resources into environmental protection; but no society is likely to choose this point. Any point on the surface OABC is efficient, but the directive might be interpreted to mean that a point on the plane OBC is chosen; in this case, the environment is totally disregarded, which is an equally improbable point in modern societies.

One of the fundamental notions which this diagram portrays is the idea of a 'trade-off'. To an economist, this means that everything has a cost, although this cost is not always equal to, or synonymous with, a cash outlay; one cannot get something for nothing. 3/ Figure 2 below is a two-dimensional cross-section from Fig. 1.

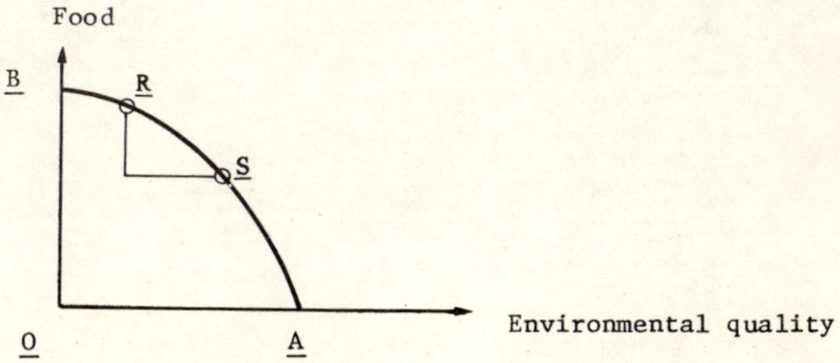

Fig. 2

Faced with the range of possible combinations of food and environmental quality shown by OAB, suppose that a society initially finds itself at Point R. A desire to increase environmental quality by moving to a point S, can only be satisfied by 'trading off' (or decreasing) food production at a rate which depends on technical factors. What this signifies is that producing a good like 'environmental quality' requires inputs of productive resources, like capital and labour, whether for scientific research, monitoring or clean up; these resources come from somewhere, and in this case, they are diverted from food production.

Recently, there has been increasing recognition of the fact that environmental quality is a desirable economic objective.4/ However, economic principles tell us that unlimited production is not possible with finite resources; this applies to material goods and to 'intangibles' like environmental quality. The development and implementation of a coastal management programme will involve countless economic decisions involving trade-offs of the kind described above. It is the job of the economist to make these trade-offs as explicit as possible, obtaining input from other scientific and technical disciplines so that a more informed and rational decision can be made.

In a single publication, it is not possible to develop a full treatment of economic principles as they apply to the management of development in the coastal environment. Readers are referred to references /10, 17 and 58/ for details. However, it is suggested that all members of the coastal management team in developing countries, be they marine scientists, engineers or sociologists, be exposed to basic economic and ecological principles which would give them all a common conceptual framework for a truly multidisciplinary treatment of the management problem.

4/ See Volume I (Summary Report) of the Workshop on Coastal Area Management in the Caribbean Region (Mexico City, 24 September - 5 October 1979), sponsored by UN/IOC/UNESCO/UNU/UNEP. Also, "The North Sea and Its Environment Use and Conflicts". Proceedings of a seminar sponsored by the North Sea Working Group (25-26 October 1979).

CHAPTER 3

Ecological Principles

A principal concern of coastal management is to understand the interrelationships among the many parts of the natural system. Ecology is the science that investigates the relationships between living organisms and their physical environment. It is a relatively new science and still in its infancy. Many of the principles of ecology are central to the development of a coastal management programme, since they provide a method for organizing information and a key to identifying the major issues we must consider when weighing the effects of a human activity against the probable response of the environment in which it will take place.

Ecosystems are the basic functioning units in nature; the term may describe the community of plants and animals living in a rotting log, or a salt marsh, or an entire estuary. The only true ecosystem is the global one and the energy it receives comes from the sun; all other terrestrial ecosystems are subsets and related to one another in varying degrees. All ecosystems are ultimately driven by solar energy, and their work is to build complex organic molecules and living organisms from simple chemicals. 1/ All ecosystems have abiotic and biotic components. Living things are further subdivided into producers (primarily green plants) that use solar energy and simple inorganic components to form complex organic compounds; consumers, that feed off the plants and other living matter; and decomposers (chiefly bacteria and fungi) that break down dead organisms into simple substances that can be reutilized by the producers.

All parts of an ecosystem are related to one another. One of the more obvious series of relationships is that found in the food web linking consumers with producers, carnivores with herbivores and the decomposers with the refertilization of the soil and water. Primary productivity is a measure of the capacity of an ecosystem to produce the green plants that form the basis of the food web. Primary productivity is extremely high in coastal ecosystems, particularly in estuaries and salt marshes. Ecosystems are also important because they can store energy. The capability of ecosystems to store energy provides stability and some insurance against boom or bust fluctuations of abundance and scarcity. For

---

1/ For a very interesting attempt to integrate economic and ecological principles through analysis of energy flows, see Howard Odum, Environment, Power and Society, New York: Wiley, 1971.

example, a saltmarsh in temperate climates may stockpile energy in the form of growing grasses during the summer. In the winter the grass dies back and is broken down into detritus that is released into the estuary where it may be an important food source. Alteration of any environmental parameter by either direct or indirect means, will disturb the ecosystems present, sometimes with dramatic results. For example, a change in agricultural practices in the coastal watershed may change freshwater drainage patterns and increase the load of silt in freshwater flowing into the sea. The increased turbidity may cut down light penetration into the water enough to decrease primary productivity. This in turn may affect the shrimps that feed on the plants and they in turn are eaten by fish that support locally important commercial fishery. 2/

One of the problems that frequently faces those involved in coastal management is that of distinguishing between natural and human-induced stresses and changes in the environment. No environment is static and natural systems are remarkably resilient to a broad range of stresses. Many natural systems are also in the process of making long-term adjustments to their environment. These adjustments may be in the form of ecological succession, a process by which plants and animals modify their environment, thus making it suitable for different species as the process advances; or, the changes may be in response to gradual long-term phenomena such as rising sea levels. It is the responsibility of the scientist to make the manager aware of such natural processes, so that he can work with them.

The recognition of significant changes in an environment is only possible if good 'baseline' information and monitoring data on that environment are available. To be meaningful the data must extend over a considerable period of time. Good baseline data may inform us which environmental variables may be classified as 'limiting factors' for that environment. Any alteration that affects a limiting factor will have profound effects. For example, it may be found that the low salinity in a coastal pond or lagoon may be limiting to many marine organisms. Species tolerant of low salinity may flourish in such an environment. Such species may include oysters that support a valuable fishery. What on the face of it may be a small change to that environment, such as a deepening of the channel into the pond to facilitate boat traffic, may increase the salinity enough to permit such predators as oyster drills and starfish to enter the pond and decimate the oyster population. A coastal management programme with a good scientific input would have served to warn the manager that such a significant change might result from a seemingly unrelated dredging project.

2/ A significant interaction of this type in a developing country context, is described in /36/.

CHAPTER 4

Requirements for a National Programme of Coastal Management

1. A National Policy for the Coastal Area

A policy may be generally defined as a (set of) guiding principle(s) or procedure(s) designed to influence the actions and decisions of individuals or groups. A fundamental requirement for a coastal management programme is a national policy for the coastal environment and the exploitation of its resources. The policy for the coastal area may, of course, be subsumed into a larger national policy covering environment and development in general.

At present, very few countries have or are formulating explicit coastal development policies. There are several reasons for this. Firstly, the global economic slowdown of the late seventies has hit developing countries very hard, and many of them are faced with worsening balances of payments and shortages of foreign and domestic capital. In this climate, short-run problems dominate the minds of policymakers, and issues of a less prominent nature such as management of the coastal area, are pushed into the background.

Secondly, there are significant cultural differences in the appreciation of environmental values which are important in formulating a policy for the coastal area. In western industrial countries, for example, environmental values are focussed on conservation and preservation of the coastal area, partly as a social and cultural reaction to rapid and uncontrolled urbanization and industrialization. The same high valuation on conservation and preservation is not felt in many developing countries, and hence, there is no pressure to develop a national coastal policy along these lines.

Thirdly, there is a widespread lack of awareness on the part of policy-makers about potential interactions between activities in the coastal area. These interactions mean that policies for individual sectors are not always independent. Thus, for example, fisheries policy and maritime policy (ports and harbours) are assumed to be unrelated. However, if the cutting of a channel through an enclosed lagoon causes a salinity change, thereby reducing local fish stocks, then a policy of facilitating coastal transport is in conflict with a policy of maximizing fisheries productivity. An overall policy is needed to guide procedures for resolving this difficulty, which is typical of the type of planning problem which arises in the coastal context. It is really up to the planners, engineers and scientists involved with technical work in the coastal area to raise the awareness of decision-makers to the need for a coastal policy; these practitioners are aware of interactions and deal with them on a recurrent basis, so they are in the best position to impress on decision-makers the real economic and financial losses associated with not having a coastal policy and management programme.

It should be noted that every policy has its costs and benefits. Suppose that a country were to make it a policy that mangrove forests in the coastal areas should not be cleared, since they serve as shoreline stabilizers and as nursery grounds for fish. But mangroves are also traditionally used as sources of timber for construction and fuel, and a village which relied on the mangrove for fuel and materials would have its way of life disrupted, and its inhabitants might have to be resettled. If this and other associated costs were deemed excessive, then an alternative policy such as management of the forest, might be chosen.

A national policy for the development and management of coastal areas should have certain characteristics. Firstly, the policy should have a significant measure of political support. The formulation of a national policy should involve wide-ranging consultations with citizens' groups, business and commercial interests, universities and research groups, and other interested parties.

Secondly, the policy must be stated in a clear, simple and intelligible form.

Finally, a national policy should be logically consistent and economically sound. Suppose that the following formulation were the national policy, "Development in the coastal area should seek to maximize environmental quality, while ensuring a rapid growth rate of production and other economic activity." This formulation, unfortunately, is neither consistent nor economically sound. A completely pristine environment would presumably represent the maximum environmental quality, and it would also exclude most human activity; maximum industrial production in the coastal area would maximize waste loadings into the air and into coastal waters, thus being inconsistent with maximum environmental quality. Residual materials could, of course, be treated to very high levels prior to their disposal into the coastal area, but the large amount of resources necessary for this task would slow growth somewhere in the economy. In fact, it is only possible to increase environmental quality by using resources diverted from other purposes; in the economist's language, there is a 'trade-off' between economic growth and environmental quality [17, 34, 58, 88]. A workable policy must recognize this relationship.

2. Definition of Management Boundaries

The coastal area has been generally defined as the zone of interaction between the land and the sea; it would thus include a landward component, the submerged lands of the continental shelf, and the superjacent waters.

Four criteria are generally used for defining the coastal area:

(a) physical criteria;

(b) administrative boundaries;

(c) arbitrary distances;

(d) selected environmental units.

In some countries, the coastal area has been defined by a combination of these criteria. Each criteria implies advantages and disadvantages and these must be weighed in the particular country context.

(a) Physical criteria. Physical criteria are commonly used for defining the coastal area. Examples of landward physical boundaries are a coastal mountain range, a watershed, or a major coastal road. The use of a watershed is sound and practical, since it is a boundary that can be readily determined and understood; it can however result in a very large land area, in the case of a large urban centre located around a bay or estuary [2]. On the seaward side, a typical physical boundary is the edge of the continental platform and the superjacent waters. The problem with this physical criterion is the fact that there is no globally recognized measurement of the continental shelf or platform; 1/ this would have to be determined by local surveys, since the global concepts are too rough.

A definition of the coastal area using physical criteria has the advantage of being simple to describe and easy to understand. Such a definition may or may not respect existing political boundaries, and it may require special legislation to be enacted. Use of the continental shelf edge as a seaward boundary will encompass most economically significant marine resources, such as fisheries and offshore oil and gas. The seaward extent may not be precisely chartable, depending on the work done by local and international hydrographic surveys.

(b) Administrative boundaries. A country may define its coastal area, particularly the landward component, by using existing political and legal subdivisions. In the German Democratic Republic, for example, the coastal area coincides with the Rostock district. The use of existing administrative boundaries has the advantage of being easily understood, readily representable, and legislatively viable. The potential drawback is that it may be too restrictive, i.e. it may not include all areas having valuable resources to be protected and managed. In the case where a country defines its coastal area using administrative boundaries, it is likely that some supplementary criteria would be used.

(c) Arbitrary distances. In the United States, the state of California has one of the most closely regulated coastal areas; the basic regulated zone is defined as extending 1,000 yards inland and 3 miles offshore, with some additional provisos. This definition, which had simplicity and ease of application in its favour, represented a political accommodation to certain interest groups which did not want to fall under the mandate of a coastal management programme. The disadvantage of using arbitrary distances for defining the coastal area is that the area may bear no relationship to coastal topography, the location of critical natural systems and the pattern of economic activity.

(d) Selected environmental units. The management area can be defined as consisting of selected environmental units, all of which are listed and enumerated. For example, in the United States, the State of Texas defined its management boundaries in terms of 'composite resource areas', viz.:

> "... nearshore Gulf (waters), inlets and tidal deltas, river-influenced bays, medium salinity bays, restricted bays, hypersaline bays and lagoons, oyster reef areas, grassflats, channel areas, submergent spoil, coastal lakes, and tidally influenced streams. The shorelands include beach and shoreface areas,

---

1/ The continental margin consists of the shelf, the slope and the rise. The transition point between the rise and the abyssal plain occurs where the slope changes from steeper than 1:1000 to flatter than 1:1000. Actual delimitation of the outer edge of the margin is easier for Pacific type margins. The precise delimitation of wide continental margins will require extensive bathymetric data. Detailed bathymetric surveys are expensive and time-consuming and have only been systematically undertaken by a small number of maritime countries.

wind-tidal flats, tidal marshes, emergent spoil, active dune areas, and active or potentially active washover channels." 2/

The definition of a management boundary using this criterion has a sound ecological and scientific basis. The environmental units themselves are not, however, easily undertood and can only be charted after extensive local survey. Legal battles may ensue over what constitutes a "tidally influenced stream", for example. More attention will be given to the question of selected environmental units in a later section.

(e) Conclusion. From the above, it is clear that no single criterion is universally applicable, nor can one criterion meet all the requirements for an effective definition of the management area. Simplicity may be the virtue of using one criterion, while completeness and environmental significance may be the virtue of another definition.

The management area, however defined, should

(1) be clearly defined in a comprehensible manner, and capable of cartographic representation;

(2) recognize, as far as possible, existing social and political subdivisions;

(3) contain resource areas of economic and environmental significance.

The problems of small islands are a special case. 3/ The area of significant interaction between land and sea can encompass the whole area of an island. In this case, coastal management and planning becomes another forum within which issues of national development planning are considered.

3. A Basic Programme of Scientific Observations on the Coastal Environment

The basic scientific data on the coastal environment required for a management programme can be broken down into the following categories, as suggested by the Joint Group of Experts on the Scientific Aspects of Marine Pollution (GESAMP) /44/:

(a) physical oceanography and meteorology;

(b) biological and chemical conditions;

(c) seabed conditions and morphology;

(d) living resources;

(e) existing discharges of sewage and industrial effluents;

(f) physical characteristics of the coast.

2/ United States of America, State of Texas. Coastal Management Programme: Hearing Draft, General Land Office, June 1976, p. 80.

3/ For an extensive treatment of these problems, see Overview on Marine and Coastal Development in the Wider Caribbean, prepared by the United Nations for the UNEP/ECLA Caribbean Environment Project (1979). See also Economic Growth and Environment Management in Smaller Caribbean Islands, Transactions of a meeting held in Barbados, 17-21 September 1979, published by U.S. Committee for UNESCO/MAB (1980).

Data in these various general categories will likely have been gathered over a certain period of time at certain locations, for example a country may have a relatively long series on meteorological data taken from seaport or airport stations. Observations on existing discharges, on the other hand, might be sporadic or non-existent. Data in all these categories can be gathered using various methods and instrumentation, implying varying cost levels. A detailed discussion of these matters can be found in reference /44/.

The programme of scientific observations and the resulting data will provide the basis for environmental monitoring of the coastal area. Environmental monitoring has two basic functions:

(1) to provide a set of benchmark or reference conditions (baselines) against which the effects of man's activities in the coastal area can be measured;

(2) to provide the basis for monitoring fundamental and long-term trends in the dynamic coastal system.

Scientific data on the coastal environment are also an input into environmental assessment studies, which are described in (5) below.

Building up an environmental database for the entire coastal area is obviously a long-term and costly undertaking. While this is being done however, all sorts of decisions will have to be made about development activities which affect the coastal environment. In these cases, certain site-specific or area-specific studies will have to be carried out and decisions made on the basis of limited information. As an example, a country planning to mine an offshore tin deposit might undertake local oceanographic, geomorphologic and biological studies to monitor the fate of disposed materials and their effects on fisheries; the required scientific studies would presumably be carried out in a limited time period, but one sufficiently long to capture seasonal variations in the environmental parameters. Decision-making on the basis of limited scientific information will be the norm for coastal management in developing countries.

In these situations the scientist has certain responsibilities. He, or she, must

(1) select the proper parameters for testing and evaluation. This means not specifying a long standard list of parameters, many of which may not be essential for the particular case;

(2) carefully collect and analyse all existing data before undertaking or recommending additional studies;

(3) specify particular sites or localities where additional sampling might be required;

(4) select the proper data collection systems;

(5) collect, evaluate and present the data in a way which makes it useful and accessible to interested scientists and decision-makers.

4. A Classification of the Coastal Area

In the large majority of developing countries, the coastal area will contain a number of distinct environmental units, such as mangroves and hypersaline lagoons. These environmental units, although different and distinct, are interconnected. In fact the pervasiveness and importance of these interconnections are among the distinguishing features of coastal management. As an example, a change in agricultural practices in the coastal watershed may change freshwater drainage patterns and increase the silt load in the freshwater flowing to the sea. The resulting increase in turbidity may cut down the light penetration enough to decrease primary productivity. This in turn may affect the shrimps that feed on the plants and which are, in turn, eaten by fish that represent an important commercial fishery. This type of problem can be avoided by (1) knowing the different environmental units, or ecosystems involved, and (2) understanding the mechanisms by which energy and material are exchanged between the ecosystems.

Thus, an essential and prominent part of every coastal management programme must be the identification of the ecosystems to be managed and an evaluation of all available environmental information on them. A list of the individual environments that require individual examination and individual attention in a management programme may include the following: 4/

1. Shoreline systems

   a. exposed shorelines; rocky shores and cliffs, sandy beaches, barrier beaches.

   b. sheltered shorelines; rocky shores and cliffs, sandy beaches, salt marshes, mud flats.

2. Estuarine systems

   large bays, tidal rivers, coastal lagoons and ponds.

3. Neritic systems

   open coastal waters and sounds.

Within any one of these units, certain features or geographical areas may be identified for preservation or conservation. In the terminology of the United States Coastal Management Act, these are called "areas of critical environmental concern". They may include breeding areas or spawning grounds, highly productive coastal wetlands or kelp beds.

Once the coastal area has been classified into its component ecosystems, a profile should be carried out for each ecosystem. A suggested outline of the general content of these ecosystem profiles appears in Table 1.

5. A Process for Evaluating Coastal Development Projects

All proposals for development in the coastal area should be subjected to a full evaluation of the net economic benefits accruing to society from the proposed project compared to various alternatives. In developing countries where manpower or finance is particularly scarce, some selection will have to be made as to which projects are screened; one criterion would be to evaluate projects above some threshold level of capital costs. For coastal development projects, a full social cost-benefit analysis must be consider two types of question:

4/ For an informative survey of coastal classification systems and a guide to the literature, see R.W. Fairbridge (Ed.), The Encyclopaedia of Geomorphology. New York: Reinhold, 1968, pp. 131-139.

TABLE 1

Outline for an Ecosystem Profile

I. A. Introduction

1. location
2. relative abundance
3. principal characteristics

B. Physical description

1. geology
   a. geomorphology
   b. topography
   c. sedimentology
2. hydrography
3. chemistry
   a. temperature and salinity distributions
   b. nutrients
   c. turbidity, heavy metals etc.

C. Major biotic components

1. primary producers
2. consumers
3. decomposers

(1–3:) major types, relative importance, diversity, adaptation to physical environment, reproduction, feeding, temporal patterns.

D. Trophic relationships

1. food chains
2. nutrient cycles

E. Relationships between physical and biotic components

F. Links between system and larger coastal ecosystems

II. A. Management considerations */

1. response to human-induced stress
2. importance to other ecosystems
3. direct and indirect values to man

B. Research needs

---

*/ major emphasis.

(1) The resource utilization question. This is in fact the focus of social cost-benefit analysis as commonly advocated. 5/ Since social economic costs and benefits differ from private pecuniary costs and benefits, and since society has a different rate of time preference from that of a private entrepreneur, the net social benefits arising from a proposed coastal development project may look quite different from the net private benefits. The point here is to determine whether society's production factors are being best used by a particular coastal development project, with 'best' being defined in terms of opportunities forgone by accepting the particular project.

(2) The environmental linkage question. This question is rarely considered in a typical social cost-benefit analysis; however, for coastal development projects, this type of question takes on special importance. The coastal area is a composite of interrelated natural systems which yield valuable resources (fish, minerals) and services (absorbtive capacity for wastes) to society. Construction and development in the coastal area can, through linkages between economic and environmental systems, adversely affect the productivity of natural systems. Any decline in natural productivity, in levels of human health and well-being, or in amenity in the coastal area are real costs associated with development projects. These costs are not considered on a commensurate level of detail with tangible economic magnitude in assessing development projects. In many industrial countries, an Environmental Impact Statement (EIS) is required of projects within and outside the coastal area. The EIS rarely attempts to quantify environmental linkages. Annexe I discusses the merits and limitations of the EIS.

A full social cost-benefit analysis of major coastal development projects should consider both the resource utilization and environmental linkage questions together. This requires negative environmental impacts of development to be translated into monetary units, since resource utilization and economic efficiency are measured in those terms. This will be a difficult problem in practice. How, for example, does one evaluate expected losses in fish catch from an offshore mining operation? While difficult, quantification of environmental costs is possible to a degree sufficient for decision-making /10, 53/.

For another example, take the case of a planner in a Caribbean island who is trying to evaluate a proposal to fill in a mangrove swamp to construct an airport extension, and assume that the proposal is technically feasible and financially profitable for the developer. The ecologist will militate against the proposal, citing the importance of the mangrove as a shoreline buffer and as a source of nutrients for young fish. 6/ The developers on the other hand will argue in favour of the proposal by citing revenues from airport taxes, increased tourist spending and local employment during construction and operation.

Firstly, the planner must determine how many of the 'benefits' cited by the developer are truly net social benefits. For example, to the extent that local construction workers are drawn away from other construction projects, their total

---

5/ For details, see United Nations Industrial Development Organisation (UNIDO), Guidelines for Project Evaluation, Vienna, 1972. (Sales no. E.72.II.B.11); Deepak Lal, "Methods of Project Analysis: A Review". World Bank Staff Occasional Papers, No. 16, 1974.

6/ As an example of the arguments, see S.C. Snedaker, "Mangroves: Their value and Perpetuation", Nature and Resources, Vol. XIV, no. 3, July/Sept. 1978, UNESCO, Paris.

wage bill cannot be legitimately counted as a net social benefit (see /17/ for a discussion). In addition, the planner must quantify the concern of the ecologist, so that ecological 'costs' associated with the airport are put on the same basis as other costs and benefits, thereby permitting an addition or summation to a grand total. What, in monetary terms, is the approximate annual value of the primary productivity of the mangrove? A complicating factor in this case is that the change is irreversible, that is once the mangrove is paved over it cannot be recreated. Then the question becomes, what is the monetary valuation placed on this piece of 'environmental capital', i.e. the mangrove?

Scientists and economists will be loathe to make this kind of hard-headed evaluation, partly because they may be accused by their colleagues of being 'unscientific'. However, this type of evaluation must be encouraged and stimulated. 7/ The point about quantifying environmental costs and benefits is not to require accuracy to three significant digits; in practice, supposedly 'concrete' financial data associated with most development projects may only be accurate to an order of magnitude.

The point is to encourage enlightened and informed discussion of the totality of costs and benefits associated with a development project in a dispassionate and analytical manner. Therefore, it is recommended that:

(1) all major coastal development proposals be evaluated in terms of their use of economic resources and their effects on the coastal environment;

(2) an agency, perhaps within the national economic planning department, be designated to carry out these evaluations, drawing upon expertise in other agencies, university departments, and in research institutions and private organizations;

(3) scientists, engineers, economists and planners be encouraged to develop a common conceptual framework and language for evaluating the wider effects of development projects on the coastal environment. This could be done through a post-graduate extension course in project evaluation, taught by a multi-disciplinary team.

6. An Institutional Mechanism for Coastal Management

Managing the coastal area requires the monitoring of a whole range of economic activity including for example, fishing, effluent disposal, oil and gas extraction and tourism; the management function will have legal, economic, scientific and political components. All this means that coastal management creates a problem for government agencies structured along classical lines.

According to the classical administrative tenets, there are four bases for organizing government agencies: (a) purpose, (b) process, (c) clientèle and

---

7/ In a number of cases, the resulting valuation on the environmental capital will be low in relation to other benefits of a project. This is because the products of the environment, say fish, may have a low unit value in relation to, say, oil; also discounting will quickly reduce benefits stemming from environmental capital over a long period of time.

(d) place. 8/ Up to the First World War, when government functions were simpler and more limited, the number of government departments was relatively small and this classical system of organization worked well. After the Second World War, a rapid increase in the number of departments took place. In recent years, government has moved into the realm of comprehensive environmental protection and management, an area not within the scope of traditional governmental functions. Modern governments "... have managed elements of the environment through public works and have been major agents of environmental change through programmes of agriculture, forestry, transportation, industrialization and nearly all of these activities subsumed under the term 'development'. ...With due regard to their particular needs and conditions, all national governments face a common problem of organization to cope with the complex, ... and evolving character of man-environment interactions." 9/

There is, however, no best way to organize government departments for development administration, and therefore no single, universally applicable mechanism for interdepartmental co-ordination required by coastal management. Governments have tended to adopt two general solutions to the problem of co-ordination in coast-related matters, viz. management by a single department and management by a multi-department agency.

Within the system of single department control, a government may choose to (1) expand the mandate of an existing department so as to encompass coastal management questions, or (2) create a new department with responsibilities for environmental matters, including coastal management. In Sri Lanka for example, the Colombo Port Commission (Coast Protection Division) has responsibility only for coastal engineering activities (hydrographic surveys, field investigations, hydraulic research) and the Government is now considering how to handle coastal management questions within the existing departmental framework. Since it would be possible to broaden the mandate of the Port Commission to include coastal management, it has been proposed that a Department of Coast Conservation be established within the Ministry of Shipping and Tourism to handle coastal management. It should be noted that the proposed solution does not, by itself, ensure communication or co-ordination between agencies with potential interests in coastal development projects. In Indonesia, three national agencies work on matters related to the coastal environment: BAPPENAS - Bureau of National Resource Management, the National Committee on Resource Inventorization; it has been proposed to create a coastal zone office within one of the above national agencies /36/.

Another solution to the problem of administrative co-ordination is to form a co-ordinating body which can oversee and delegate particular functions to existing agencies. An example of this type of arrangement is found in the proposed Environmental Protection Council in Trinidad and Tobago. This body, made up of representatives of existing departments, 10/ academia, industry and citizen groups has some interesting features.

8/ Examples of each are (a) defence, housing; (b) finance, taxation; (c) labour; (d) a river basin commission, or an authority for a depressed area.

9/ United Nations, 1974. Organization and Administrative of Environmental Programme. (E.74.II.H.5), p. 18.

10/ Including the Departments of Health, Planning and Development, Industry and Commerce, Petroleum and Mines, Local Government and Finance, the Attorney-General's Office, and the Water and Sewerage Authority.

The creation of specialized coastal management agencies, patterned after river basin authorities, is not likely to be an effective alternative for nationwide coastal management, but this type of special purpose agency may be useful for management of a limited geographical area, e.g. a bay around which a large, urban-industrial complex has developed.

Whichever solution is taken to the institutional problem, certain guidelines should be satisfied:

(1) the agency should consider the whole broad set of problems of economic development, resource use and environmental protection in the coastal area;

(2) the agency should ensure the active participation of representatives from all relevant government departments, with an interest in coastal development;

(3) the agency should have a high-level executive group;

(4) the agency should not interfere with the local planning process, except to raise attention to local projects which have regional or national implications;

(5) the agency should have all large-scale coastal construction projects and engineering works submitted to it for consideration at the feasibility stage, and it should undertake or delegate responsibility for project evaluation studies;

(6) the agency should have a management information branch or a unit which undertakes to assemble and present relevant data to it;

(7) the agency should have active participation from local universities, institutes and research laboratories;

(8) the agency should have active participation from industrial associations and public interest groups;

(9) the agency should report regularly on its activities, and it should disseminate news of its deliberations widely.

7. Commissioning of Baseline Studies

"Virtually all coastal developments lead to changes in topography and in geographical configuration of the coastline. Such changes lead inadvertently or by design to alterations in coastal processes." 11/ If changes in the physical and chemical characteristics of the coastal area, with consequent effects on the biota, are to be evaluated, it is essential to have a description of the initial state of the coastal ecosystem, that is, a 'baseline'. Any scientific study which attempts to measure certain parameters for this purpose is called a 'baseline study'.

11/ M. Waldichuk, "Observations in the Marine Environment Related to Estuarine Ecosystem Disruption in Coastal Area Development", prepared for GESAMP Working Group on the Pollution Implications of Sea-Bed Exploitation and Coastal Area Development. (London, 10-14 January 1977), 1977.

Baseline studies are undertaken as part of the process of environmental management, and they have as their general objective the protection of human health and coastal fisheries. Baseline studies can be undertaken in connexion with any coastal development (dredging, filling, port construction, etc.), but the emphasis here will be on baseline studies preceding offshore oil exploitation. There are several good reasons for this emphasis.

Worldwide production of crude petroleum from offshore areas amounted to 484.737×10^6 metric tons (1974), of which 77.6 per cent came from developing countries. 12/ Of the total offshore production by developing countries, almost 50 per cent came from the Middle East, slightly over 30 per cent came from Caribbean countries, about 13 per cent came from Africa, and some 6 per cent from developing countries in the Far East. Production of offshore petroleum will continue to grow by all accounts, and in value terms it will dominate all other coastal activities in many developing states. Furthermore, the potential harm to health and living resources is of great concern to policy-makers in both developed and developing countries.

8. ANNEXE I: The Environmental Impact Statement (EIS) as a Control Device For Coastal Management

The environmental impact statement (EIS) is a legal and administrative tool for providing decision-makers with a basis for determining the desirability of any proposed man-made modification of the environment. An EIS may be required from a small-scale project, such as the damming of a stream, to a large-scale project, such as overflights by a supersonic airliner. The EIS has been widely advocated as a control measure in coastal management.

There is no standard format for an EIS, and there are no standards for the analytical methods used. 13/ However, most published statements to date consist of:

(a) a lengthy description of the coastal environment in which the proposed project will take place;

(b) a lengthy exposition of the likely impact of the project on the environment based on a review of published literature from other areas;

(c) a justification of the project, discussing likely adverse impacts and weighing them against anticipated benefits.

The early versions of the EIS depended heavily on published literature, and specific original research was almost never undertaken.

An example of a modern EIS is the Final Environmental Statement for the Proposed OCS Oil and Gas Lease Sale in Offshore Southern California. 14/ The description of the physical and economic environment goes on for over 800 pages;

12/ United Nations, World Energy Supplies, 1950-1974, (Statistical Series J, no. 19), 1976.

13/ R.B. Ditton and T.I. Goodale (Eds.), Environmental Impact: Analysis Philosophy and Methods. University of Wisconsin Sea Grant Programme:(Madison, Wisconsin) /WISC-SG-72-111/, 1972.

14/ Prepared by the Bureau of Land Management, U.S. Department of the Interior. The EIS is in 5 volumes, totals more than 3300 pages and weighs more than 4 kg.

the environmental impact of the proposed sale covers some 450 pages, alternatives to the proposed lease sale cover some 160 pages, and one entire volume is devoted to legal and political co-ordination of agency activity. Irrespective of the analytical content and substantive merits of such a document, it seems well beyond average human capacity to read, retain and understand such a staggering amount of diverse data, facts and opinions. This phenomenon is sometimes called 'information overload'. The result is often that a project is approved simply from weariness on the part of the decision-makers. The methodological basis for the EIS is also subject to criticism. According to one scientist familiar with European experience in environmental problems, "The EIS may be appropriate for political or public relations purposes but it is unsatisfactory as a scientific exercise." 15/

Another very serious shortcoming with the application of the EIS has been the tendency to file an EIS after the planning and decision-making for a particular project; thus, the EIS serves only to justify decisions which are already in the process of being implemented, instead of being an integral part of the project planning. This tendency is exacerbated when an EIS does not consider several alternatives to a particular project. 16/

There is no doubt that the use of EIS has contributed to improved decision-making in respect of coastal resource use in industrialized countries. While the process has permitted effective advocacy of environmental protection however, this improvement has come at a considerable cost in additional government staff, consultant expenditure and interagency conflict and co-ordination. The fact that an EIS is prepared for a proposed coastal development does not, of itself, ensure that a better decision will be made than would occur in the absence of an EIS. Secondly, an EIS should be an integral part of the economic evaluation of a project.

15/ "Pollution Control - Environmental Impact Statements or Critical Path Approach?", Marine Pollution Bulletin, Pergamon Press, August 1979.

16/ In the case of a proposal for outer continental shelf (OCS) petroleum development, one alternative might be OCS production and another might be increased imports.

CHAPTER 5

Special Topics in Coastal Management

So far the fundamental elements and functions required within a national coastal management programme have been discussed. Within each of these there are a large number of detailed technical issues which cannot be effectively treated in a single publication. Therefore, a selection of topics based on both importance and interest must be made. The emphasis below is on the most important principles, and the reader is referred to the literature for details.

1. Setting Environmental Standards in the Coastal Area

One of the three categories of coastal problems identified earlier was the pollution of coastal and estuarine waters. 'Pollution' requires the identification of the polluting substance and a measured concentration above the acceptable baseline level of the substance. This section will discuss some problems in the formulation of water-quality criteria for coastal areas, although the principles apply to the setting of other environmental standards, such as air quality standards.

It is necessary to maintain or improve coastal water quality because:

(a) the biological productivity of the coastal area is a function of several variables, including water quality;

(b) marked declines in water quality can seriously threaten human health when people contract water-borne diseases by direct contact or by eating contaminated fish and shellfish;

(c) a decline in water quality reduces the economic value of the water for beneficial uses.

Water quality criteria are defined as "the required scientific information on which a decision or judgement may be based concerning the suitability of the environment to support a desired use, recognizing that the health of man is paramount and that the latter can be affected either directly or indirectly." 1/

1/ GESAMP, Reports and Studies, No. 5 (Rome, FAO, 1976), p. 1.

There are two basic approaches to the control of water quality. The first approach is to define effluent standards. This means that all discharges into coastal waters by economic units are required to meet certain physical or biochemical criteria. To take a simple example, an electric power plant may not be allowed to discharge its cooling water into the coastal area if it is above a certain temperature, the threshold temperature being the 'criteria' or the 'standard'. The control of water quality at the point of discharge is the method used by the United States Environment Protection Agency. Very complex and detailed standards are required on an industry-by-industry basis. It also requires knowledge of alternative treatment technologies and their associated cost functions for producers. 2/ Finally, it requires a large administrative and legal apparatus for monitoring, control and litigation.

The second possible approach is to define water quality standards for the receiving waters themselves; that is to say, coastal waters shall always, for example, have dissolved oxygen concentrations of x parts per million. From the point of view of economic theory, this has something to recommend it. Producing units in the coastal area are now free to meet this standard in a cost-minimizing way, using the technology of pollution abatement and the absorptive capacities of the waters up to the standard. This approach is administratively and technically simpler to implement. However, there is a strong possibility that this approach will carry a high economic cost, if the standards for the receiving waters are set arbitrarily high and if care is not taken in determining when the standard is exceeded. 3/ An effective system of water quality control can be established by combining elements of the two above approaches. 4/

A model water-quality criteria system would, in principle (1) apply to all coastal waters, including estuaries, lagoons and marshes, and (2) include a complete set of quantitative exposure-effect relationships covering all toxic agents, effects and environmental variables. 5/

Unfortunately, it is not likely that a quantitative exposure-response relationship of the kind advocated by scientific groups can be derived for even one substance entering the coastal environment, given the current state of knowledge. Secondly, in large coastal countries such as India, Australia, Brazil etc., it is simply impractical to formulate water quality criteria for all coastal receiving waters and for all toxic agents. Therefore, a national programme of water quality control in coastal areas must represent a scientifically sound, economically sensible and administratively practical compromise from the scientific ideal.

2/ For an illustration and discussion, see the article by Spofford et. al., in /58/.

3/ This point is discussed below.

4/ A combined system is described in B.T. Hart, "Establishment of Quality Criteria for Water Use in Australia", United Nations Water Conference, E/CONF.70/TP.109, (25 Nov. 1976).

5/ The methodological ideal for a water quality control programme is described in GESAMP, "Principles for Developing Coastal Water Quality Criteria", Reports and Studies, No. 5 (Rome; FAO, 1976), p.3.

A water quality programme for the coastal area, including the formulation of both criteria and standards, should satisfy the following guidelines:

(1) it should clearly state the areal coverage of the programme;

(2) water quality criteria should be formulated, as a first priority, for the most critical receiving bodies and for the most sensitive areas of the coastal environment;

(3) there should be a baseline review of discharges into coastal waters and existing water-quality conditions;

(4) there should be a statement of the beneficial uses of coastal waters to be protected by the programme;

(5) water-quality standards should be set in such a way as to minimize the cost of achieving the desired quality with a specified degree of certainty;

(6) in formulating water-quality standards and programmes, developing coastal states should begin by reviewing the international literature, particularly that of scientific groups like GESAMP.

It was mentioned that when setting standards for receiving waters, administrative simplicity might be associated with a high economic cost. Figure 3 below illustrates one way in which this result might be obtained. 6/

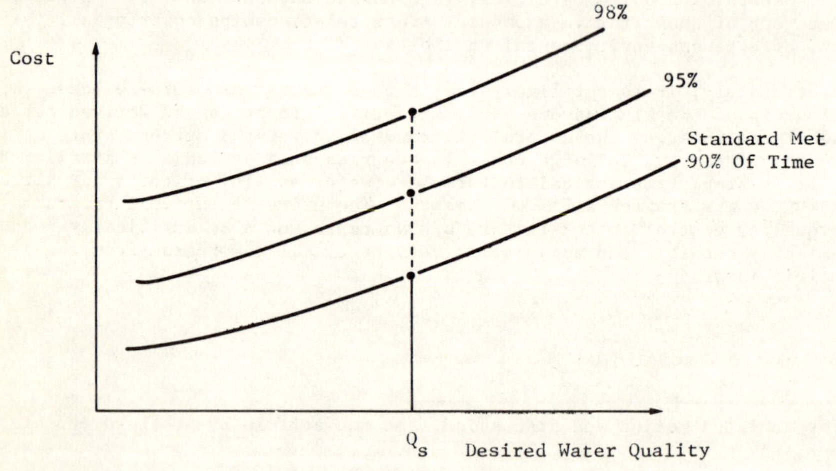

Fig. 3

6/ Following B.T. Bower and A.V. Kneese, Managing Water Quality: Economics, Technology, Institutions, Baltimore, Maryland: Johns Hopkins University Press, 1968.

Treatment cost, which is one component of the environmental cost of water policy, is an increasing function of desired water quality. Water quality is a random variable, being affected by seasonal climatic variations. Therefore, one measurement at one point in time is insufficient to determine that the waters are polluted and probabilistic system of measurement should be used. From Fig. 3, for a given level of water quality, a stringent policy of meeting the standard 95 per cent of the time is much more costly than a somewhat lenient policy where the water quality standard must be met only in 90 cases out of 100. Yet the benefits associated with the stricter policy will, in many cases, not be significantly greater than the benefits associated with the less strict policy. Therefore, in setting water quality standards, effluent standards and industrial licensing policies, it is essential that an economic cost-benefit approach be applied to the evaluation of alternative programmes.

Water quality criteria and standards are required data for an integral part of the environmental assessment and project appraisal process. Thus, water quality criteria and standards must be available to the project-review group for large-scale coastal developments. Monographs on water quality management, with provision for the regular up-dating of criteria, should be issued by the appropriate government body for the use of planning bodies, such as the coastal project-review group.

2. Water Quality Modelling

Mathematical models (whether expressed as simple nomographs or as complex computer simulations) are being increasingly used in water quality and environmental investigations to express the dynamic response of the system. The primary purpose of many of these models is to simulate interactions among various physical and chemical constituents of the environment. In this context mathematical models refer to all quantitative techniques that are used to analyse these relationships but here we will consider only relationships between water flow (movement) and water quality.

The resolution of a model refers to the degree of model complexity and the accuracy of the description it provides. Low resolution models are generally simple mathematical relationships, require relatively little input data and can be solved relatively quickly and easily. High resolution models utilize complex mathematical relationships, require large amounts of input data, and are usually solved with digital computers.

Most water quality models applicable to the coastal zone consist of two sub-models: (a) the hydrodynamic sub-model, and (b) the water quality sub-model. These two sub-models are linked to provide an integrated response sensitive to both system hydrodynamics and water quality changes.

The effects of the hydrodynamic sub-model and the water quality model are combined with a linking equation. The most commonly used linkage is the mass balance equation, which can be expressed in one dimension as follows:

in which C_i is the concentration of the constituent, x is the space dimension, V is velocity, A is area, D is the dispersion coefficient, S represents transport of mass across system boundaries, R represents biochemical transformations within the system, and t represents time. Time and spatial dimensions are the independent variables in the system. The constituent concentration C_i is the dependent variable being simulated in time and space by the model. If more than one constituent (pollutant) is being modelled, an equation similar to the above is written for each one and all equations are solved simultaneously. The hydrodynamics of the system are linked to the equation through the A, V and D coefficients. The S coefficient allows for mass transfer across the boundaries. The biological and chemical reactions are linked to the equation through the R coefficient. All coefficients may be time or space variables, or all the coefficients may be replaced by more complex terms with additional coefficients. These terms may be expressed as functions of any of all dependent variables being modelled. It seems clear the equation(s) may be modified to become either very complex and difficult to solve or relatively simple, depending upon the needs of the application.

Estuaries present the most complex conditions for hydrodynamic modelling. For the purpose at hand consideration will be restricted to one-dimensional sub-models of estuaries.

Hydrodynamic sub-models. Two distinctly different types of time scales are used in one-dimensional estuary models. One includes velocity variation caused by tidal action directly in the model, thus making pollutant distribution a function of tidal elevations. The second type of time-scale includes only average net water movements over a period of time equal to or greater than one tidal cycle. Since intertidal velocities are excluded in this case, these effects on the concentration distribution are considered to be represented by the dispersion coefficient D.

There are available both analytical (or exact) methods as well as numerical methods for solving the one-dimensional form of the mass balance equation. One form of exact solution is based on the assumption that equilibrium conditions exist in the estuary and that all parameters are constant with time and distance. Another anlytical solution to the mass balance equation is based on the assumption that D and V vary as algebraic functions of time. The major limitation of the approach is that the cross-sectional area of the channel is considered uniform throughout the length of the estuary.

Due to irregular geometry and unsteady flows in estuaries, the mass balance equation does not permit analytical solution in many cases. Numerical methods have sometimes been used to obtain greater freedom in varying the parameters of the equation.

Water-quality sub-models. There are several water-quality parameters considered to be of importance for water quality sub-models. These include dissolved oxygen (DO), biochemical oxygen demand (BOD), algae, nitrification, temperature and sediment transport. Fairly sophisticated methodologies have been developed and applied to the evaluation of the above parameters. However, not all will be considered herein.

Dissolved oxygen (DO) is considered to be the most important coastal zone water quality parameter . Because water contains less than 1 per cent by volume of oxygen at normal temperatures, the entire community of aquatic organisms in the coastal zone is critically sensitive to the oxygen available to them.

The dissolved oxygen concentration is the net result of oxygen fluxes into and out of the water. Fluxes into the water are primarily from atmospheric re-aeration across the water surface and oxygen production by photosynthesis.

Oxygen leaves the water primarily due to consumption by bacteria during the oxidation of organic substrates (BOD) and ammonia (nitrification), and by respiration of aquatic organisms.

Although photosynthesis generates considerable amounts of oxygen under the same conditions, it is restricted primarily to daylight hours. During night hours aquatic plants consume oxygen and release carbon dioxide into the water. The result is a diurnal cycle of DO and CO_2 in a water body which is rich in vegetation. Water quality sub-models for simulating the effects of photosynthesis are available.

In streams the re-aeration process across the surface involves physical absorption of oxygen from the atmosphere.

Biochemical oxygen demand (BOD) is customarily defined as the amount of oxygen required by bacteria as they stabilize decomposable organic matter under aerobic conditions. It has been found from experience that a reasonably large percentage of the biochemical oxygen demand is exerted in five days at 20° C. Thus the five day BOD has become a standard value. Although many models have been used, in practice applications to natural waters have involved a first-order reaction introduced by Streeter and Phelps as early as 1925. In this equation the scale of the reaction is proportional to the amount of oxidizable organic matter left at any time.

Techniques for the modelling of several types of aquatic systems have developed very rapidly in recent years. The literature on this subject is too great to be adequately reviewed herein. However, Patten (1968) has provided a reasonably good review of plankton models. 7/ The models include the effects of light, temperature, and nutrients, and are restricted primarily to linear functions which can be relatively easily solved.

For reasonably simple low resolution water quality models, the system of nomographs developed by the U.S. Environmental Protection Agency is recommended. 8/ The techniques embodied in the above report provide good estimates for instream concentrations of BOD, ammonia, and DO resulting from various loading conditions. Charts and tables are available in the report to assist in estimating the hydraulic characteristics of a wide range of flow regimes, including estuaries. In spite of the relatively low resolution of these techniques, they are recommended for situations where the constraints of time and money preclude more sophisticated approaches.

7/ B.C. Patten, "Mathematical Models of Plankton Production", Internationale Revue der Gesampten Hydrobiolgie, 53:357, 1968.

8/ U.S. Environmental Protection Agency. Simplified Mathematical Modeling of Water Quality, report prepared by Hydroscience, Inc., Washington, D.C., 1971.

CHAPTER 6

The Role of Regional Programmes in Coastal Management and Development

Development and management of the coastal area, including the land and sea, are fundamentally national concerns. At the same time, there are a number of international efforts operating at the level of the 'marine region' or 'regional sea', which are relevant and important for coastal management in developing countries. 1/

A 'regional sea' or 'semi-enclosed sea' might be defined as "... having a water area of at least 50,000 square nautical miles, at least fifty percent of its circumference being occupied by land, and that it be a primary sea, rather than an arm of a larger semi-enclosed water body. The sea should be bordered by two or more States in order to be regional in nature." 2/ The Caribbean Sea and the Gulf of Guinea are two examples. Perhaps the most prominent example of an international organization structuring its efforts on this concept is the United Nations Environment Programme (UNEP), which has a special Regional Seas Programme.

In economic terms, regional programmes relating to development and management of the coastal environment have certain clear justifications. Where nations border the same regional sea, no nation can independently pursue a policy of marine waste disposal without impinging on the environmental health of one or more other nations. Clearly, a rationalization and co-ordination of environmental policies would be required. Secondly, migratory living resources require international co-operation to optimize the sustainable yield and allocate it among countries. Finally, these programmes can help to economize on personnel and equipment costs for countries carrying out the types of surveys and resource inventories discussed in this publication. For all these reasons, regional programmes such as those carried out by UNEP, are very important for developing coastal states.

1/ For a survey of regional arrangements in marine affairs, within and outside the United Nations system, see L.M. Alexander. "Regional Co-operation in Marine Science", Report prepared for the Intergovernmental Oceanographic Commission (IOC) of UNESCO, United Nations (Ocean Economics and Technology Branch, and the Food and Agriculture Organization of the United Nations.

2/ Ibid. p. 1-12.

Participation in regional programmes encompassing coastal management and development should be seen as being parallel and supportive to the fundamental national programme. 3/ Regional programmes in coastal management also offer possibilities of technical co-operation among developing countries (TCDC), particularly in the training of personnel in various pure and applied sciences relevant to coastal management.

3/ For a specific example of these general points in a developing region, see "Marine and Coastal Area Development in the Wider Caribbean Area: Overview Study", (E/CEPAL/PROY.3/L.INF.13, 25 October 1979). Submitted to a meeting of Government-Nominated Experts to Review the Draft Action Plan for the Wider Caribbean; prepared for UNEP by UN/DIESA.

CHAPTER 7

Conclusion

The integrated management of the coastal area and its resources requires the interlocking of many activites, such as scientific research, data gathering, policy analysis and economic evaluation, in an organized manner. It has been the aim of this part of this publication to give specificity to this general concept in terms of what is required for a national programme.

Developing a management programme takes time and money. The many possible programme options render it meaningless to say that a developing country must have "x coastal managers, y scientists, and spend z amount of dollars to develop a programme". Some countries have begun the process of developing a coastal management programme by a short-term consultant's study, while others have spent millions of dollars over several years. Some countries have created special institutions to act as focal points for a coastal management programme; where others have left the institutional and administrative responsibility more diffuse.

Whichever option a developing country chooses, the programme must have certain basic inputs from policy-makers, scientists, engineers, economists, and the general public; these have been described in this part. The programme must also carry out basic analytical evaluations of development alternatives, presenting to decision-makers the consequences of these alternatives for the economy and for the environment. This too has been described in some detail. The character of these elements will be basically invariant under any chosen programme option, although their levels may differ.

There is much to be gained from a cross-fertilization among the disciplines involved in coastal management. For example, scientists traditionally take no interest in economics and finance; yet, an understanding of national economic issues can help the scientist to understand the background against which decisions are taken, which otherwise might seem completely arbitrary to him or her. Similarly, the planner with some knowledge of the principles of coastal engineering may better appreciate the potentially serious consequences of faulty siting and construction in the coastal area. It is strongly recommended that all staff in the coastal management programme be exposed, even if somewhat generally, to the multi-disciplinary character of management in the coastal area.

References

1. Armstrong, J.A., et al., Coastal Zone Management: the Process of Programme Development, Sandwich (USA), Coastal Zone Management Institute, 1974.

2. Australia, Ministry for Conservation, The Westernport Bay Environmental Study (1973-1974), Melbourne, Ministry for Conservation, 1975.

3. Baker, J.M. (ed.), Marine Ecology and Oil Pollution, New York, Halsted Press, 1976.

4. Baram, M.S., Environmental Law and the Siting of Facilities: Issues in Land Use and Coastal Zone Management, Cambridge (USA), Ballinger, 1976.

5. Bartlett, C.A. and Moulder, D.S. (eds.), A Bibliography on Marine and Estuarine Pesticide Pollution, Marine Biological Association of the United Kingdom, Marine Information Centre, Plymouth, 1976.

6. Baumol, W.J. and Oates, W.E., "The Instruments for Environmental Policy", in E.S. Mills (ed.), Economic Analysis of Environmental Problems, New York, Columbia University Press, 1975, pp. 95-128.

7. Bishop, Bruce, "An Approach to Evaluating Environmental, Social and Economic Factors in Water Resources Planning ", Water Resources Bulletin, 8 (4): 724-734, (1972).

8. Bjorkland, M., "Achievements in Marine Conservation, I. Marine Parks", Environmental Conservation, 1 (3): 205-223, (1974).

9. Bower, B.T., "Studies of Residuals Management in Industry", in E.S. Mills (ed.) Economic Analysis of Environmental Problems, 1975, pp. 275-320.

10. Chacko, E., "Environmental Assessment Techniques: Their Applicability to Coastal Management", paper presented to Workshop on Coastal Area Management in the Caribbean Region, Mexico City, 24 September-5 October 1979.

11. Clark, J., Coastal Ecosystems Management: A Technical Manual for Conservation of Coastal Resources, New York, Wiley, 1977.

12. Clark, M.J., "The Relationship between Coastal-Zone Management and Offshore Economic Development", Maritime Policy and Management, 4 (6): 431-449, (1977).

13. Conservation Foundation, Rookery Bay Land Use Studies, Vols. I-X, Washington, D.C., 1975.

14. Cowell, E.B. (ed.), Ecological Effects of Oil Pollution on Littoral Communities, London, Applied Science, 1971.

15. Cronin, L.E. (ed.), Estuarine Research: Geology and Engineering, New York, Academic Press, 1975.

16. Cullen, P.W., "Management of the Coastline of Port Phillip Bay", in Australian Institute of Engineers, the Engineer, the Coast, and the Ocean. Proceedings of the 2nd Conference on Coastal and Ocean Engineering, 1975, pp. 106-112.

17. Devanney, J.W., Parable Beach: A Primer in Coastal Zone Economics, Cambridge (USA) MIT Press, 1977.

18. Devanney, J.W., "The OCS Petroleum Pie", Massachusetts Institute of Technology Sea Grant Programme, Report No. 75-10, 1975.

19. Devanney, J.W., "Key Issues in Offshore Oil", Technology Review, 76 (3) 21-25, January 1974. Edited at the Massachusetts Institute of Technology.

20. di Castri, F. and Mooney, H.A. (eds.), Mediterranean Type Ecosystems: Origin and Structure, New York, Spring-Verlag, 1973.

21. Ditton, R.B., et. al., Coastal Resources Management, Lexington (Mass.), D.C. Heath, 1977.

22. Ditton, R.B. and Stephens, M., Coastal Recreation: A Handbook for Planners and Managers, Washington, D.C., National Oceanic and Atmospheric Administration, 1976.

23. Dolan, R., Hayden, B. and Heywood, J., "A New Photogrammetric Method for Determining Shoreline Erosion", Coastal Engineering, Vol. 2, pp. 21-39 (1978)

24. Duane, D.B., "Sedimentation and Coastal Engineering: Beaches and Harbours", in Stanley and Swift (eds.), Marine Sediment Transport and Environmental Management, New York, 1976, pp. 493-517.

25. Duane, D.B., "Sediment and Coastal Engineering: Placer Mineral Resources", in Stanley and Swift (eds.), ibid., pp. 535-556.

26. Economic Commission for Europe, "Ecological Aspects of Development Planning", ENV/SEM.2/3 (30 September 1975).

27. Economic Commission for Europe, "Proceedings of the Seminar on National and Regional Planning as Framework for Local Planning", HBP/SEM.7/4.

28. Efrat, Elisha, "Physical Planning Aspects of Israel's Coastal Strips", Ministry of the Interior, Planning Department, Jerusalem, 1973.

29. Efrat, Elisha, "Outline Scheme of Israeli Coastal Strips", Ministry of the Interior, Planning Department, Jerusalem, 1974.

30. "Environmental Impact Statements: An Analysis of Six Years' Experience by Seventy Federal Agencies", Council on Environmental Quality, Washington, D.C. 1976.

31. Fallah, M.H. and Stark, R.M., "Movement of Spilled Oil at Sea: a Literature Review", Marine Technology Society Journal, 10 (1): 3-18, (1976).

32. Fisher, W.L. et al., Environmental Geologic Atlas of the Texas Coastal Zone, Galveston-Houston, Bureau of Economic Geology, Austin, 1972.

33. Flawn, P.T., Environmental Geology: Conservation, Land-use Planning and Resource Management, New York, Harper and Row, 1970.

34. Freeman, A.M., Haveman, R.H. and Kneese, A.V., The Economics of Environmental Policy, New York, Wiley, 1973.

35. Goldberg, E.D. (ed.) Strategies for Marine Pollution Monitoring, New York, Wiley, 1976.

36. Hanson, A.J. and Koesoebiono, "Settling Coastal Swamplands in Sumatra: A Case Study for Integrated Resource Management", Bogor Agricultural University, Bogor (Indonesia), 1979.

37. Hard, G.B. and Palmer, H.D., "Sedimentation and Coastal Engineering: Ocean Dumping", in Swift and Stanley (eds.), Marine Sediment Transport and Environmental Management, New York, 1976, pp. 557-577.

38. Hayden, B. and Dolan, R., "Classification of Coastal Environments of the World", Department of Environmental Sciences, University of Virginia, Charlottesville, 1976.

39. Hey, R.D. and Davies, T.D., (eds.), Science, Technology and Environmental Management, London, Saxon House, 1976.

40. Inhaber, Herbert, Environmental Indices, New York, Wiley, 1976.

41. International Union for Conservation of Nature and Natural Resources. "Ecological Guidelines for Development in Tropical Forest Areas of South-East Asia", Morges, Switzerland. IUCN Occasional Paper No. 10, 1974.

42. Isard, Walter, Ecologic-economic Analysis for Regional Development, Glencoe, Free Press, 1972.

43. Jacobsen, N.K., "The Development of Sedimentary Coastal Systems", paper presented to First European Ecological Symposium: Ecological Processes in Coastal Environments, University of East Anglia (U.K.), 1977.

44. Joint Group of Experts on the Scientific Aspects of Marine Pollution, Report of the Working Group on Marine Pollution Implications of Coastal Area Development. XIth Session, 25 February-1 March 1980, GESAMP XI/5.

45. Joint Group of Experts on the Scientific Aspects of Marine Pollution, Report of the Working Group on the Impact of Oil on the Marine Environment: Second Session. Rome, FAO/UNEP, 1976 (GESAMP/VIII/4).

46. Joint Group of Experts on the Scientific Aspects of Marine Pollution. Report of the Working Group on the Impact of Oil on the Marine Environment. Rome, FAO/UNEP, 1976 (GESAMP VIII/4/1-2).

47. Joint Group of Experts on the Scientific Aspects of Marine Pollution. Report of the Working Group on the Principles for Developing Coastal Water Quality Criteria. Rome, 1976 (GESAMP/VIII/6).

48. Keyes, D.L., "Land Development and the Natural Environment: Estimating Impacts", Washington, D.C., the Urban Institute, 1976.

49. King, C.A.M., Beaches and Coasts, New York, St. Martin's Press, 1972.

50. Kneese, A.V., Ayres, R.Y. and d'Arge, R.C., Economics and Environment, Washington, D.C., Resources for the Future, 1970

51. Krutilla, J.V. and Fisher, A.C., The Economics of Natural Environments: Studies in the Valuation of Commodity and Amenity Resources, New York, Johns Hopkins Press, 1975.

52. Macaulay, H.H. and Yandle, T.B., "An Economic Evaluation of Water Quality Management Systems", Water Resources Research Institute, Report No. 58, Clemson, South Carolina (USA), 1976.

53. Mäler, K. and Wyzga, R.E., Economic Measurement of Environmental Damage: A Technical Handbook, Organization for Economic Co-operation and Development, Paris, 1976.

54. Massachusetts Institute of Technology, Offshore Oil Task Group, The Georges Bank Petroleum Study, 3 Vols., Cambridge, 1973.

55. McEachern, John and Towle, E.I., "Ecological Guidelines for Island Development", International Union for Conservation of Nature and Natural Resources, Morges, Switzerland. Publication No. 30, 1974.

56. McRoy, C.P., "Seagrass Ecosystems: Research Recommendations of the International Seagrass Workshop", Australian Marine Science Bulletin, 46: 8-13, (1973).

57. Meyer, J.R., "Setting Environmental Standards: An Economist's View", New York, National Bureau of Economic Research, 1973.

58. Mills, E.S. (ed.), Economic Analysis of Environmental Problems, New York, Columbia University Press, 1975.

59. Moore, S.F., "Offshore Oil Spills and the Marine Environment", Technology Review, 78 (4): 60-67, (1976).

60. Moore, S.F. and Dwyer, R.L., "Effects of Oil on Marine Organisms: A Critical Assessment of Published Data", Water Research, 8: 819-827, (1974).

61. Nelson-Smith, A., Oil Pollution and Marine Ecology, London, Elek Science, 1972.

62. Nihoul, J.C.J. (ed.), Modelling of Marine Systems, Amsterdam, Elsevier, 1976.

63. Nisbet, Ian C.T., "Air Pollution vs. Food Production: An Illogical Trade", Technology Review, 76 (3):28, (1974).

64. Odum, W.E., "Ecological Guidelines for Tropical Coastal Development" International Union for the Conservation of Nature and Natural Resources, Morges (Switzerland), 1976.

65. Organization for Economic Co-operation and Development, Review of Marine Science and Technology: Ireland, Paris, 1975

66. Palmer, H.D., "Sedimentation and Coastal Engineering: Structures", in Stanley and Swift (eds.), Marine Sediment Transport and Environmental Management, pp. 519-534.

67. Pearson, E.A. and De Fraja Frangipane, E. (eds.), Marine Pollution and Marine Waste Disposal, Oxford, Pergamon Press, 1975.

68. Pedrick, Jr., L., "Land Use Control in the Coastal Zone: the Delaware Example", Coastal Zone Management Journal, 2 (4): 345-368, (1976).

69. Phillips, P.W., "Analysis of Decision Making in Coast Protection", in R.D. Hey and T.D. Davies (eds.), Science, Technology and Environmental Management, London, Saxon House, 1976, pp. 47-48.

70. Ranwell, D.S., "Strategies for the Management of Coastal Systems", paper presented to First European Ecological Symposium: Ecological Processes in Coastal Environments, University of East Anglia (U.K.), 1977.

71. Ray, G.C. and Sprunt, A., Parks and Conservation in the Turks and Caicos Islands, report prepared for Hon. R.E. Wainwright, Administrator, Turks and Caicos Islands, 1975.

72. Reed, A.W., Ocean Waste Disposal Practices, Park Ridge, New Jersey, Noyes Data Corporation, 1975.

73. Russell, C.S., Residuals Management in Industry: A Case Study of Petroleum Refining, Baltimore, Johns Hopkins University Press, 1973.

74. Sacchi, C.F., "Italian Littoral Lagoons", paper presented to First European Ecological Symposium: Ecological Processes in Coastal Environments, University of East Anglia (U.K.) 1977.

75. Schlesinger B. and Daetz, D., "A Conceptual Framework for Applying Environment Environmental Assessment Matrix Techniques", Journal of Environmental Sciences 16 (4): 11-16, (1973).

76. Scottish Development Department, "North Sea Oil and Gas: Coastal Planning Guidelines", Aberdeen, 1974.

77. Sedgwick, R. and Arthur, D.R., "A Natural Pollution Experiment: The Effects of a Sewage Strike on the Fauna of the Thames Estuary", Environmental Pollution, 11(2), 2: 137-159 (1976).

78. Shepard, F.P. and H.R. Wanless, Our Changing Coastlines, New York, McGraw-Hill, 1971

79. Simeon, Paul, Défense et aménagement du littoral, Paris, Bureau Central d' Etudes pour les Equipements d'Outre Mer (BCEOM), 1972.

80. Society of Environmental Engineers, London, Environmental Engineering: Its Role in Society. Six Vols. Proceedings of a symposium held at Imperial College of Science and Technology, London (19-21 April 1966).

81. Sorenson, J.C., "A Framework for Identification and Control of Resource Degradation and Conflict in the Multiple Use of the Coastal Zone." Unpublished Master's Dissertation, Department of Landscape Architecture,

University of California (Berkeley), 1971.

82. Stanley, D.J. and Swift, D.J.P. (eds.), Marine Sediment Transport and Environmental Management, New York, Wiley, 1976.

83. Steers, J.A., "Protection des zones côtières", Strasbourg, Council of Europe 1974.

84. Steward, R.J., "Oil Spills and Offshore Petroleum", Technology Review, 78 (4): 46-59, (1976).

85. Swift, D.J.P., "Coastal sedimentation", in Stanley and Swift (eds.), op. cit.

86. Thomas, W.A. (ed.), Indicators of Environmental Quality, New York, Plenum Press, 1972.

87. Tihansky, D.P. and Meade, N.F., "Economic Contribution of Commercial Fisheries in Valuing U.S. Estuaries", Coastal Zone Management Journal, 4 (2): 411-421, (1976).

88. Tolley, G.S., "The Resource Allocation Effects of Environmental Policies", in E.S. Mills (ed.), Economic Analysis of Environmental Problems, pp. 133-163.

89. Traves, N.H., Skeats, R.S., and Ellis, P.L., "Coastal Management Investigation, South-East Queensland", in Australian Institute of Engineers, The Engineer, the Coast and the Ocean. Proceedings of the 2nd Conference on Coastal and Ocean Engineering, 1975, pp. 98-105.

90. United Nations, Economic Commission for Europe, "Environmental Impact Assessment", ENV/R.53 (23 December 1976).

91. United States Senate, Committee on Commerce, North Sea Oil and Gas: Impact of Development on the Coastal Zone, Report prepared for 93rd Congress, 2nd Session, Washington, D.C., 1974.

92. United States Army Corps of Engineers, The National Shoreline Study, 4 Vols., Washington, D.C., 1971

93. United States Department of the Interior, Ad Hoc Committee for Design of Baseline Studies. "Guidelines for the Design and Conduct of OCS Oil and Gas Development Environmental Baseline Studies", Washington, D.C., 1976.

94. Vagners, J. and Mar, P., Oil on Puget Sound: An Interdisciplinary Study in Systems Engineering, Seattle, University of Washington Press, 1972.

95. van der Maarel, E., "Environmental Management of Coastal Dunes in the Netherlands", paper presented at the First European Ecological Symposium: Ecological Processes in Coastal Environments, University of East Anglia (U.K) 1977.

96. Victor, P.A., Pollution, Economy and the Environment, London, Allen and Unwin, 1972.

97. Watt, K.E.F., Ecology and Resource Management, New York, McGraw-Hill, 1968.

98. Westhoff, V. and Schouten, M.G.C., "The Structure and Diversity of European Coastal Systems", paper presented to First European Ecological Symposium: Ecological Processes in Coastal Environments, University of East Anglia (U.K.), 1977.

99. Weyl, P.K., "Pollution Susceptibility: An Environmental Parameter for Coastal Zone Management", Coastal Zone Management Journal, 2 (4): 327-343, (1976).

100. Whipple, Jr., W., The Delaware Estuary System, Environmental Impacts and Socio-economic Effects: Environmental Quality and Its Evaluation, Washington, D.C., National Technical Information Service, 1975.

101. World Health Organization, Guides and Criteria for Recreational Quality of Beaches and Coastal Waters, Geneva, 1975.

102. Zenkovitch, V.P., Processes of Coastal Development (translated from Russian by O.G. Fry and edited by J.A. Steers), Edinburgh, 1967.

PART II

Approaches, Problems and Guidelines

CHAPTER 8

Part II: Introduction

In recent years, there has been an increasing interest in coastal area management. This is perhaps in recognition of two important groups of factors. The first relates to the characteristics of the coastal area and the second to the seaward expansion of the resource jurisdiction of coastal states.

Because the coastal area is where land, sea and air meet and interreact to influence the whole environment, any development activity in the coastal area requires consideration of a much broader spectrum of elements (e.g. geological and environmental) and their interrelationships hitherto excluded under the scope of the existing framework. While coastal areas are ideal locations for industrial activities for obvious reasons of access, supply, market and waste disposal into the sea, the special characteristics of the coastal environment may constitute drawbacks which must be taken fully into account.

To disregard these characteristics could result in the undermining of the very objectives intended to be achieved. Moreoever, the coastal area is biologically rich and aesthetically pleasing. It contains the large majority of population, possesses some of the highest-grade agricultural land, accounts for the lion's share of the tourist trade, and is the growth pole of expanding economic activity. There is therefore a great demand for the use of the coastal area, resulting in competitive claims and conflict of uses. In most countries, the coastal areas are subject to numerous single-purpose laws executed by different agencies and institutions. Hardly any horizontal links exist between these laws and among their executing agencies. This situation often results in lacunae, conflicts or overlapping jurisdictions.

Given the special characteristics of coastal areas, it seens that a broader spectrum of elements should be taken into account in making a development decision. There is a need to identify these elements so as to ensure rational decision-making processes, avoid costly alteration at a later stage, and reduce possible adverse effects which may prove to be irreparable. 1/

The concept of 'coastal area management' has undergone certain changes. When first introduced in the United States by the Coastal Zone Management Act of 1972,

---

1/ For further elaboration, see Chapter 16.

one of its major concerns was to protect the rather unique environmental position of the coastal area which is especially sensitive to the adverse consequences of congested human settlement, i.e. to pollution and the reduction of amenities caused by urban sprawl and industrial development. There was therefore the need for better planning and regulation of diversified activities impinging on the coastal area. Although the need for wiser use of natural resources was recognized, protection of the coastal area predominated in the concept. As time has gone by, this concept of coastal area management appears to have been considerably broadened, in terms both of substance and scope of application, as a result of practices in the United States and various other countries.

As will be seen, different types of legislation have now been introduced in various countries to deal with their specific needs in the coastal areas. Some legislative measures are designed for planning purposes, e.g. regulations on zoning and the imposition of building prohibitions or the acquisition of land, as methods of controlling urban development in coastal areas. Some are designed to deal with critical situations in certain sections of the coast, particularly because of mineral exploitation or tourism. Others have as their goals conservation, preservation or protection in coastal areas. Laws to establish nature reserves and sanctuaries, or to prevent oil pollution are examples of this type. Still others are designed to provide financial and institutional needs, either requiring better co-ordination among existing administrative units or establishing a central co-ordinating unit to ensure consistent implementation of decisions by all agencies concerned.

It should be pointed out however that the legislative measures so far taken specifically for the development of coastal resources remain scanty. Where such measures have been taken, they are largely single-purpose or piecemeal. A systematic approach has yet to emerge. In those countries where some form of resource planning or management exists, such planning or management is usually with reference to a narrow strip along the shoreline and coastal waters. This is the case, for example, in Cyprus (with 50 metres of highwater marks), Mauritius (from the coral reef to the shoreline of high watermarks within 1 km.), the United States (within three miles of the territorial sea) and Togo (from the high tide within the landwards area of 100 m).

The coastal areas in most countries remain governed by the general legal regime applicable to all areas. They are subject to the same laws governing land use, town and country planning, and resource exploitation. They have not been treated differently or separately. Consequently, the general legal regime may not have taken into account adequately the special characteristics of the coastal areas. In addition, there is a need for providing horizontal links between the numerous single-purpose laws executed by different agencies and institutions.

The second group of factors has led to an increasing interest in coastal area management which relates to the seaward expansion of the resource jurisdiction of coastal states. In the last few years, more than 90 states have claimed some form of marine resource jurisdiction (generally a 200 nautical mile zone adjacent to their land mass). These zones, taken as a whole, contain a significant part of the sea's total living and mineral resources. It is estimated that over 90 per cent of the total volume of the world commercial fish catch is taken from coastal and nearshore areas. These areas also include practically all known and presently exploitable mineral resources, all of the world's proven reserves of offshore hydrocarbons and most of those having potential economic value in the next decade. 2/ The exercise of resource jurisdiction over the ocean will necessitate the adoption of managing measures, as for example mapping, boundary delimitation, resource management, traffic and pollution control. It is of paramount importance, therefore, that each coastal state establish some kind of marine resource policy and management programme to provide a system leading to proper planning, and permit conscious and informed choices among development alternatives so as to promote sound utilization of resources and enjoyment of amenities. Furthermore, supplies of material or technology for resource development in the ocean have to come from the coastal area, and any impact resulting from the development - be it economic, social or environmental - will also fall primarily on the coasts; the coastal area is therefore the ideal focal point for managing the uses of not only that part of the land near the sea, but also the water and marine resources offshore, as well as the interaction between the two.

Some coastal states will need to prepare maps for their territorial seas, economic zones and continental shelf, and to determine boundaries with their neighbours. This will not be an easy task. Information indicates that many coastal states parties to the 1958 Geneva Conventions have yet to delimit their territorial sea boundaries, offshore boundary agreements only existing in some countries. 3/ Delimitations pertaining to the continental shelf are almost non-existent since they are generally further seaward. While the absence of some boundary agreements might have been the result of political difficulties, the technical complexities involved must not be underestimated. Now that the new limits are even further seaward and the areas included are far larger than before, the task becomes correspondingly more difficult.

Coastal states may wish to conduct various activities in the ocean area, such as to erect installations for the production of energy from the water, sea currents and winds in the zone, provided that they do not interfere with the use of recognized sea lanes essential to international navigation. A coastal state may therefore need to prepare laws and regulations in respect of maritime traffic, navigational aids, cables and pipelines, artificial islands, prevention of pollution, scientific research and customs, and immigration, all of which are issues associated with the exercise of jurisdiction over the extended resource area. Where necessary, sea lanes and traffic separation schemes may also be needed for regulating in particular the passage of oil tankers, and ships carrying dangerous substances. While these laws and regulations may already exist in some countries, they must be re-examined in the light of the extended jurisdictions, particularly from the point of view of enforcement. In other countries, they must first be established.

Resource management is probably the most important concern. With regard to living resources, a coastal state will need to determine for example the total catch allowable without depleting the stock and its capacity to harvest, particularly if foreign fishing vessels are to be allowed to share part of the resources. This affords the opportunity for a rationalization of the fishing

---

2/ "Economic Significance in Terms of Sea-bed Mineral Resources of the Various Limits Proposed for National Jurisdiction", Report of the Secretary-General, A/AC.138/87, 4 June, pp. 37-38.

3/ United Nations Legislative Series on National Legislation and Treaties Relating to the Territorial Sea, the Contiguous Zone, the Continental Shelf, the High Seas and the Fishing and Conservation of the Living Resources of the Sea. United Nations Publications 1974.

effort, so as to enable fishing to proceed at the level of the maximum sustainable yield. It may be possible to achieve the same catch level with a reduced fishing effort; it is, for example, estimated that the same or a higher catch could be obtained from the North Sea using half the vessels at present employed in fishing there. 4/ Some coastal states will also need to deal with such matters as licensing, fishing vessels and equipment, species and quotas to be caught, seasons and areas of fishing, the type and size of fishing gear that may be used, training of personnel, transfer of technology, enforcement procedures, and conservation of living resources.

As for mineral resources, even greater technical difficulties may be envisaged. Being a new field, exploitation can be a very costly undertaking. Some of the basic problems relate to the identification and assessment of resources, the accessibility of exploration data, methods of exploitation, and problems of the environmental impact of exploitation and beneficiation. Each of these items will require scientific study and investigation, and the adoption of a wide spectrum of measures. The emerging new law of the sea is likely to resolve the doubts which formerly existed concerning the extent of coastal states' rights over the mineral resources of the continental shelf. Thus for example, petroleum and gas in the outer portions of the margin at a depth of 600-10,000 ft. which might be commercially exploitable, would be clearly recognized as belonging to the coastal states. Licences granted by such states, and investment and operation under them, would then be secured.

In view of the increasing interest in coastal area management and development, it would appear that countries either in the process of preparing their legislation or unaware of the need for coastal area management would be interested in having information on: the kind of problems existing in other countries, the approaches and solutions to these problems found in those countries and how legislation has been used to achieve the objectives. Moreover, because coastal area management is a new discipline, existing literature is not only very limited but deals mostly with national or local concerns. There is therefore a need at the international level for collecting and disseminating information on the relevant legislative and policy measures taken in different countries. It is in recognition of this particular need that the present study has been prepared.

In order to take into account all relevant legislative measures and national approaches, this study uses the term 'management' to embrace a broad spectrum of concerns and interests, including planning, resource utilization, conservation, preservation, and administrative needs. For the same reason, a broad definition of the term 'coastal area' is employed, that is the seaward limit to coincide with the extended resource jurisdiction, and the landward boundary to include not only the narrow strip of the shore, the intertidal areas, salt marshes, wetlands, lagoons or river estuaries, but also the adjacent areas where the impact of a given activity may be felt or where the cause for concern originates. Under such a broad concept of coastal area management, the scope of activities which could be termed 'coastal' and the coverage of laws described by the word 'legislative' are inevitably voluminous. In order to keep this study within manageable size, a selection has been made on the basis of comparative importance, relevance and representativeness.

It should be noted here that certain basic issues in coastal area management are commonly shared among developing countries. Firstly, in contrast to the fact

---

4/ A.V. Lowe, "The New Law of the Sea Opportunities for Ocean Industries", Ocean Management, Vol. 4, 1979, p. 152.

that private enterprises play an important role in developed countries, it is the government agencies in developing countries which play a central role in practically all important development activities. These agencies are staffed politically by the leaders who spearhead the development. Hence, coastal activities, whether exploitation or conservation of resources, have to be channelled through government circles. Even within government, the central or state government can always overrule the organs of local government and give directives to these organs, whether municipal or commercial, as to how each is to use its powers. This result tends sometimes to stifle initiative and increase dependence upon the higher authority.

Secondly, many countries in Europe and North America seem to favour the co-operation of public and private agencies on a scale not hitherto envisaged. Local authorities in these countries have been given a new range of powers with financial support. Much emphasis is now laid on the importance of public participation and it is now accepted that members of the public have a right to be consulted about, and to take part in, the formulation of local planning policies. The habit of setting up public enquiries or commissions to consider controversial matters of public concern has yet to take root in developing countries.

But on the other hand, if the governments of developing nations wish to move aggressively into management of their coastal zones, it appears they already enjoy a good deal of the requisite legal authority. The governments of many of these countries own the mineral deposits and other natural resources while in others, the State has significant legal control over resources exploitation. Water is considered public property in large areas of the hemisphere; beaches are public in many countries and private owners of ocean-front property are subject to important restrictions. Construction of hotels and development of tourism is under governmental supervision in most of these nations. If proper management of the coastal zones necessitates expropriation of certain private properties, many developing nations are in a position to do so more easily and less expensively than would be the case, as for example in the United States, where private property may be taken only through complicated legal proceedings and at prohibitively high costs. 5/

It would appear then that the concept of coastal management can be implemented under existing legislation and for many countries, within the development plan structure already established by their town and country planning acts. However, successful management in coastal areas requires also that the controls available under planning legislation be supplemented by using such techniques as zoning or selection of sites. For those areas which are subject to intense pressures, additional measures may also be needed. The urgent task is to identify these areas, their needs and the inadequacy of existing legislation.

5/ Payment of the fair market value of such property renders government efforts on any scale very expensive. The relevant laws in developing countries, for example in some Latin American nations, require only "prior indemnification", leaving the issue of value open. The Mexican Constitution provides property may be taken for "public utility" and the amount of compensation to be paid will be "the tax assessed value". Tax assessed value is usually less than fair market value. The Venezuelan Constitution provides that in case of the taking of land for reasons of "serious national interest", payment may be deferred for a specified time period or may be made partially in government bonds. Similar patterns exist in a number of Latin American countries.

The principal need in many developing countries is perhaps to establish a legal scheme pulling together various administering agencies and the fragmented laws and regulations applicable to coastal activities. The Venezuelan Council on Renewable Natural Resources and the Norwegian Natural Resources Law may prove a useful scaffold upon which a coastal area management scheme could be built. 6/

Another serious issue is that while many laws and statutes look like perfect models, it would be a mistake to suppose that they are all put into effect in their entirety. When it is a question of their implementation, various factors are involved. Very often lack of funds, equipment or personnel impedes execution. Almost by definition, developing nations suffer from the lack of a management framework and a shortage of funds, trained technicians and managers to execute programmes. These are very important issues and they must be resolved.

It may be said that the management of coastal resources in developing countries faces problems similar to many coastal areas of the developed countries. But social conditions and economic pressures in these developing economies necessitate different management perspectives and techniques. Caution must be exercised in adopting methodologies, legislation or standards from other countries' attempts at coastal management. For developing countries a major element must be the development of new concepts which should recognize the sound foundations of some traditional management frameworks while exploring new methods and procedures for expanding and integrating resource control links.

6/ See Chapter 11.

PART II

SECTION ONE

Critical Issues and Planning Aspects in Selected Coastal Areas

CHAPTER 9

Some Country Examples

i. Bahamas

In the Bahamas, development plans are being prepared on an island-by-island basis, treating the coastal area as a separate planning entity. 1/ On the basis of a recent land resource survey, priorities of uses within the coastal area have been assigned to recreation, tourism, harbours, residential development, preservation, and conservation. Zoning orders are issued to cover all areas of outstanding natural beauty, historical and archaeological sites, scenic vistas, parks, and recreation areas. Planning guidelines for the control of land use and development incorporating the classical treatment of land use, have recently been published. 2/ While these guidelines are applicable to areas both inland and shoreland, special rules are established for the coastal area. The setback requirement for building in the coastal area is not fixed by limit but by requiring "a view of the sea", whereas in the city the limit is set by distance (from 15 to 30 ft) from the street. 3/

ii. Barbados

Barbados is the most densely populated island in the Caribbean. This is an important background factor in any consideration of the current and potential use of the coastal areas, as one of the island's two major industries, tourism (the other is sugar) relies heavily on the beaches as a natural asset and is becoming a major land user of the coastal area, and the distribution of resident population has become increasingly concentrated along the south and west coasts.

The overall physical structure emerging from the trends in population movement (and tourist development) is of a coastal linear corridor of varying width, extending on either side of Bridgetown southeast almost to Seawell Airport and northwards to Speightstown, a distance of about 20 miles. Within this corridor lies considerable scope for future infill and growth, and provided care is taken not to overload the infrastructure, particularly with roads (already under some

---

1/ Communication from the Commonwealth of the Bahamas, dated 18 February 1976.

2/ See Planning Guidelines for the Control of Land Use and Development in the Commonwealth of the Bahamas, published by the Government of the Bahamas (no date).

3/ ibid., Sections 8.2 and 9.19.

congestion) and beaches, it may well be that the best strategy is to concentrate further growth here. It would have the advantage of leaving substantial areas of coastline relatively undisturbed and secondly, would not encroach on the good inland agricultural areas.

Despite the rapid extension of urbanization along the south and west coasts there still remain substantial areas around the north and east coasts of unspoilt natural scenery, as well as areas of ecological significance. The distinction between areas of ecological importance which have to be preserved without change in their natural state, and areas of natural beauty where varying degrees of limited and strictly controlled development can take place, is an important one in the context of a crowded island with the twin pressures of tourism and population affecting the use of coastal areas. Each area has to be examined carefully to determine what must be conserved and what impact different types of change can have on the natural habitat.

Barbados' main natural asset is its fine white beaches. No mineral natural resources have yet been discovered and the island's other natural endowments are basically geographical ones.

Historically, Barbados was well placed to become a port and a thriving centre for commerce. The combined effects of the Northeast Trade winds and the convergence of the north and south equatorial currents make it the most convenient first land-fall for sailing ships approaching the Caribbean from Europe. It has had the further advantage of being south and east of the main hurricane strike areas in the region.

For tourism, the island has a natural advantage in climate as a result of the cooling breezes of the trade winds, which make it somewhat cooler than other islands in the region. Moreover, there are clearly recognizable wet and dry seasons, and the time of year when skies tend to be clearest and the cooling trade winds most pleasant, corresponds with the North American and European winters. The island is also well placed geographically to receive tourists from South America.

A further natural asset is the reef surrounding the island which is not only the natural habitat for fish, but also ensures shark-free beaches.

One unfortunate aspect of the fishing industry has been the small-scale dynamiting of the reef. Although prohibited by law since 1904, this has continued spasmodically until quite recently. It not only destroys fish life and therefore affects inshore reef catches, but affects the growth of coral. An important decision has therefore been taken to protect a 2 mile stretch, 1/4 mile from the shore of the best and deepest coral, by establishing a marine underwater park off Holetown where there will be complete restriction of any activity likely to damage underwater flora and fauna.

The chief problem relating specifically to coastal areas has been the continuing practice in the fishing villages and other small settlements (often adjacent to tourist development) of throwing garbage, including dead animals, into the sea. The problem was exacerbated until recently by infrequent and unreliable garbage collection services and also the prohibition of burning rubbish on the beach.

The need here is for a programme of environmental health education to emphasize the role of the sea as a resource rather than as a utility.

A number of bodies are involved in the implementation of policies relating to coastal areas, but there is no systematic collection of data. Neither is there any real co-ordination. There is quite considerable scope for consultation, but co-ordination tends to be restricted to financial control, with decisions on projects made on an individual basis and the initiative for bringing up projects resting with an individual ministry or department, which decides its own priorities. As a consequence, the link between comprehensive development planning and implementation and management is rather tenuous.

Comprehensive development planning is vitally necessary so that the balance between (i) the use of natural resources, (ii) infrastructure and (iii) the carrying capacity for urban development of coastal areas is maintained. As far as coastal areas are concerned, (i) the beaches (natural resource), (ii) water supply and sewage disposal (infrastructure) and (iii) the amount of urban/ tourist development, their environmental capacity has not yet been exceeded with the exception of the roads on south and west coasts. Nevertheless, the situation needs to be constantly monitored and correlated to the present and proposed development of the area concerned.

In particular, the maintenance of high-quality water supply is of paramount importance and will require integration of a water resource development plan with the overall physical and economic planning of the island. Protection of the purity of water supply is intimately connected with methods of sewage disposal. Thus it could be that water supply and sewage should come under one autonomous body rather than be divided between the Ministry of Health and the Waterworks Department.

Similarly, it is essential that an effective stormwater drainage system takes account of the ultimate development plan for the area it is to drain, and this would require the co-ordination of the various infrastructure projects for coastal areas including the Bridgetown Harbour Extension, the Bridgetown Sewerage Scheme and major road proposals.

What these examples point to is the need to have properly integrated data collection in related subject areas, perhaps best correlated in a physical development plan which would be an ongoing and cyclical process of monitoring and review.

iii. Cyprus

The Town and Country Planning Law of Cyprus 4/ requires the Minister of Finance to carry out a survey of the Republic and submit to the Council of Ministers for approval an "Island Plan". 5/ The Plan is to indicate the general policy to be followed in promoting and controlling development, and specifically the location of population, industry and commerce, tourism, public transportation and conservation of areas of special social, historic, architectural, cultural interest or natural beauty. The Island Plan is subject to constant review. Under the same Law, the Minister may also be requested to prepare "local plans" for

---

4/ Town and Country Planning Law, 1972 (English translation prepared at the Ministry of Justice), Nicosia, published by the Government, No. 90.

5/ *ibid.*, Part III, Section 7.

specific areas or detailed proposals for any part of a local plan. 6/ In addition there are detailed regulations governing streets and the construction and alteration of buildings in the country and in coastal areas. 7/ Any "development" (which includes the carrying out of buidling, engineering, mining or other operations in, on, over or under immovable property) must first obtain a development order (i.e. planning permission). 8/

iv. Eastern European Countries

Coastal area development in Eastern Europe is inseparable from the concept of state ownership and national economic planning. In these countries, the basic means of production, the banks, and the building and transport sectors are state-owned. Under the Constitution of the German Democratic Republic for example, mineral resources, mines, power stations, barrages and large bodies of water, resources on the continental shelf, industrial enterprises, bank and insurance companies, nationally owned farms, traffic routes, railway transport, ocean shipping, civil aviation, post and telecommunications are either nationally-owned property or activities which are organized by the State. In Bulgaria and the U.S.S.R., water resources fall in the same category. Bulgaria and Yugoslavia consider fish and other sea animals to be state property. Thus coastal area development activities have to be carried out largely at state level. Government authorities decide the most appropriate use and development of natural resources in coastal areas, in the light of the strategy for social and economic development adopted at a given period for the whole country.

Generally, the national economic plan is drawn up on the basis of a long-term projection (for a period of 25 to 30 years), a short-term projection (usually for 5 years concerning the growth and distribution of productive resources), and regional development requirements which include aspects of local planning. Regional plans provide the necessary framework for dealing with such issues as allocation of industrial and agricultural enterprises, cities and urban centres, transport, communications, water supply, sewerage, protection of nature, improving working and sanitary conditions, and recreation facilities.

In the Soviet Union, the users of a given resource are required to keep the necessary record showing the quantitative and qualitative utilization. 9/ Land utilization is based on the land cadastre as seen from the point of view of quality of land, and the changes occurring on the land surface. It is obligatory to keep cadastres for all natural resources. 10/ Poland has a new law on the planning of industrial and residential development 11/, the purposes of which are to ensure proper development of particular regions of the country, to establish a proper relationship between areas of production and of service facilities, and to establish the best conditions for development and for the protection of natural resources and values. Plans are made at the national, regional and local levels. The Planning Commission is responsible for making national and regional plans.

6/ ibid., Part IV, Section 10.

7/ The Streets and Buildings Regulation Law (incorporating amendments made by laws passed up to 31 December 1973) Nicosia, February 1973.

8/ Town and Country Planning Law, 1972, ibid., Section 20(1).

9/ For RSFSR, that is the Law of 29 October 1960. Vedomosti Verhovnogo Soveta RSFSR, 1960, No. 40.

10/ By virtue of the Law of 29 October 1960.

11/ Law on Planning of Industrial and Residential Development, 31 January 1971.

The Ministry of Administration, Regional Development and Protection of the Environment is responsible for local plans. All this applies to the planning and development of the coastal area. The use of land in seaports and in the coastal belt is subject to the Law on Grounds Administration in Towns and Urban Settlements 12/; in each case of development, special consultation with the appropriate authority of sea administration and National Councils is required. Building, construction, bridges, tunnels, roads, railroads, reservoirs and industrial plants are regulated by the Building Law. 13/ This Law requires that the architectural shape of any building should harmonize with the neighbourhood and landscape and should improve the beauty of the surroundings. Buildings or industrial plants must not affect the water, air, soil or landscape quality established by the State. The conversion of green terrains to other uses must be justified for social or economic reasons, capital gain alone being insufficient.

The development of tourism in Poland's coastal regions is contained in the general comprehensive plan of tourist industry development up to 1990. 14/ The long-range development programme for coastal areas takes into account facilities necessary for both domestic and foreign tourists, mainly from Czechoslovakia, the German Democratic Republic, Scandinavia and the U.S.S.R. The entire Polish coastal area is treated as one of the five zones divided for tourist development purposes. The fundamental conditions of the implementation of a tourism development programme in coastal regions include (i) the maintenance of the natural and cultural heritage of the country; (ii) the development of tourist infrastructure and industry; (iii) the creation of an economic and financial basis for the implementation of the programme; and (iv) implementation of the programme by effective institutional means. A number of coastal cities including Gdansk and Szcecin are considered very attractive for tourists. The Old Town in Gdansk has been restored and plans have been made to build a number of modern hotels and motels in the coastal region and also to develop transport for tourist purposes.

Romania has a law on spatial planning for the whole territory to ensure optimal land utilization. 15/ Any emplacement on land must be preceded by special emplacement studies which take into account established guidelines and environmental requirements. Any activity dealing with mining, industry, energy, roads, agriculture or tourism must follow this procedure and be approved first by the local spatial planning Commission and then by the State Committee of Planification. Special legal regimes are established for different categories of land, e.g. agricultural, forests, and areas designated for industrial and building purposes. The use of land for mining, industries and recreation is also subject to special decrees prescribing various environmental requirements. 16/

12/ Law on 14 July 1961 on Grounds Administration in Towns and Urban Settlements, Dziennik Ustaw, No. 21, Item 159, 1961.

13/ Building Law, 24 October 1974.

14/ Plan of Tourist Industry Development of Poland until 1990 prepared in April 1973 by the Institute of Tourism for the Planning Commission at the Council of Ministers. Besides the general plan, regional plans were also drawn up which in turn include local plans worked out by local and town planning bodies. The local plans deal with such issues as beaches, use of dwelling houses in holiday resorts, tourist development of coastal areas and development of sports and water navigation.

15/ Law on Spatial Planning of Territory, Town and Rural localities, 29 October 1974.

16/ See for example, the Law on Environmental Protection 1973; Decree of Council of Ministers No. 311, 1963 on water pollution; Decree No. 541 1965 on Sanitation.

In Yugoslavia, a new spatial organization has been initiated in the Adriatic region. Efforts are being made to link this area to the hinterland, so as to bring this region within the framework of Yugoslavia and the Mediterranean. The approach adopted is to integrate settlements in existing urban areas as far as possible, open new settlements as an integral part of the overall plan, improve existing settlements and repair the damaged environment.

Bulgaria, the German Democratic Republic, Poland, the Soviet Union and Czechoslovakia all have legislation regarding the protection of water, air, soil and forests, construction, space planning, urbanization, recreation and tourism, all of which are equally applicable to the coastal regions. The protection schemes cover maintenance, restitution and rational exploitation of all natural resources.

Existing legislation in most eastern European States imposes obligations on enterprises whose activities may affect the environment, and particularly that of the coast. They must ensure proper and rational use of the environment and its resources and minimize damage to it. They are required to install necessary equipment for the protection of the water, air, soil and forests and to introduce technological changes in their plants to preserve environmental quality. Expenditures for treating pollutants and developing facilities for protecting environmental quality are regarded as necessary production costs, and as such, are subject to deduction. In these countries, special regional and urban planning plays an important role in the protection, development and rehabilitation of the environment. The degree of integration of that planning into the general economic planning is however different in each country.

v. Ecuador

Information is lacking on the existence of ground water in the coastal areas of the provinces of Guayas and Manabí.

Coastal waters are polluted in many places. For example, discharge from the city sewer of Esmeraldas is causing pollution up to six miles from the coast, having adverse effects on the recreation activities on the adjacent beaches. In Puerto de la Libertad area, pollution was found up to 50 km north of the peninsula (Pacoa). The Puerto de Guayaquil suffers from bilge discharge from boats and untreated industrial effluents, e.g. from food factories, brewery, detergent and soap factories affecting the Gulf. Pollution in Punta Carnero and Manta areas is due to unloading coastal shipping vessels and construction work near the seawall.

Although the population density in the coastal area is not high, there is a severe unemployment problem, possibly owing to poor management and lack of new development plans.

An integrated approach to coastal area development could be adopted, but it is not considered desirable to adopt general guidelines, in view of the fact that conditions in the coastal areas and the continental shelf vary according to their location.

vi. El Salvador

Approximately 80 per cent of the essential information base is lacking here. There is not enough high-quality information covering the coast (continental portion) and no information on the offshore area. The result of these deficiencies is that any analyses covering the sea coast and offshore area are merely extrapolations, leaving a wide margin for error.

In view of the fact that all critical meteorological situations - those which give rise to storms followed by floods - originate in disturbances in the Pacific

Ocean, it is clear that the strengthening of activities to obtain basic information concerning the continental coastal area and sea region is a matter of the utmost priority. Better planning is required and there is a need for overall regulation.

Floods occur in June, September and occasionally in October. Studies have shown that during those months, the disturbance of the Pacific, the so-called Inter-tropical Convergence Zone (ITCZ), is very close to the coast. The main problem lies in predicting well in advance the arrival of the disturbance which will enter the region, so that potential floods can be anticipated at a sufficiently early stage. At present, the arrival of the disturbance is perceived only at the time when its effects begin to be noticeable in the form of prolonged bouts of rainfall. The flood warnings therefore do not reach the population early enough.

The floods in the coastal area relate to the fact that only 8 per cent of the country area, excluding the coffee-growing portion, is naturally wooded, and these woods are thin and scattered.

There exist no particular laws or regulations governing shoreland use, zoning or beach access. However, the Forestry Act governs the conservation, improvement, restoration and growth, exploitation and rational management of the national woods and forests.

In view of the relative natural homogeneity of the coastal strip and its short length, it would appear that there is a need for an integrated approach to coastal area development. But there remains an even greater need for basic information and appropriate studies.

vii. Ethiopia

The coastal area of Ethiopia, situated in the northern part of the country, comprises a large sandy plain having a width of 10 to 70 km and extending for a total coastal length of about 1000 km. For water resource studies and eventual development purposes, the coastal area, known as the Red Sea catchment area, constitutes one of fourteen drainage basins of the country and ranks eleventh in area with about 44,000 km^2. The two Ethiopian ports of Massawa and Assab are located along this coastline.

With the exception of these two ports however, most of the other parts of the coastal area are sparsely populated, mainly with nomads and semi-nomads who depend for their living on cattle, goat and camel raising. Along some river courses there are settlements that depend on irrigation from groundwater. Some farmers in these localities annually produce two crops of vegetables, melons and field crops by taking advantage of the summer floods originating in the adjacent highlands and the winter rains of the coastal plains. Besides moistening the soil, the floods deposit silt that greatly improves the soil's fertility and texture.

So far no systematic investigations of the resources of the coastal area have been undertaken with the exception of offshore and limited shoreland explorations for petroleum, potash exploration in the Danakil Depression, located just to the south of the coastal area, and geothermal investigations in the Rift Valley area and around the port of Massawa. These efforts will need to be supplemented by further studies before definite development plans can be drawn up.

The land and water resources of the coastal area can support sound agricultural development as shown by some of the ongoing irrigation schemes in the area. There are a number of torrential rivers flowing down the escarpments of the southern and western highlands. Huge fresh groundwater resources are also believed to exis exist. However, more data and information need to be compiled and carefully analysed prior to any meaningful development. At present, the National Water Resources Agency of the Ethiopian Government has plans to set up hydrometeorological stations to carry out general surveys to enable a proper assessment of the land and water resources potential and to lay down a sound basis for future development opportunities in the coastal area.

By and large, there is a lack of an adequate information base and the various natural resources are at different stages of investigation and development. For systematic development of the coastal area, more comprehensive planning is required; the present piecemeal and ad hoc approach to development will need to be carefully analysed and regulated so as not to hamper future profitable investment opportunities.

In some of the developed localities of the coastal area, such as the ports of Assab and Massawa and the village of Tiho, a chronic shortage of fresh water has been observed. Ground and surface water development can play an important role in relieving this shortage provided systematic studies are carried out. It would appear that groundwater exploitation close to the coast may carry the risk of salinity.

The torrential rivers of the coastal area flow during the summer months of July, August and September, which constitute the rainy season of the adjacent highlands. Regulation of the resulting floods has proved profitable for agricultural development in some places. Uncontrolled, however, these floods have been the cause of hazards in other places. These torrential rivers have over the years seriously eroded the highlands, which are now in dire need of an afforestation programme to arrest further erosion, but they also have carried large volumes of silt which are deposited further downstream. This silt has greatly improved the soil's fertility and texture in the lower lands where profitable agricultural development is possible. On the other hand, the few dams that have already been constructed, such as those at Zula and Ghinda, have experienced serious sediment encroachment upon their reservoir capacities. As a result of this, the Zula dam is now no longer operational, as its reservoir has practically been filled with silt. Hence, there is a need for proper watershed management and careful measurement of sedimentation prior to development.

The ports of Massawa and Assab are the only places where significant industrial and commercial activities exist. Both of them have their own municipalities and town master plans. The town of Massawa, with a population of about 25,000, is well planned, and has a cement factory, salt processing works, fisheries, several small industries and workshops and a modest commercial centre providing good town facilities and amenities. The town of Assab, having a population of some 20,000 has, among other things, an oil refinery, salt processing works and fisheries. These two ports do not yet have any port development master plans. Although, at present, port facilities and equipment appear to be adequate to handle the existing volume of traffic, there is a need for a comprehensive port development master plan covering a period of about 25 years into the future. In fact, with this in view, basic statistical data and pertinent information have been compiled and analysed and detailed terms of reference prepared to enable the carrying out of the necessary master plan study for both ports. It is foreseen that this study will give serious attention to the need for co-ordinating port development plans with town planning and with the inland transport approaches to the ports.

At present, due to the small size of the port towns coupled with the under-developed resources of the coastal area, sea pollution does not pose a serious problem, and as such may not appear to merit urgent attention. However low the level of sea pollution may be, the main sources of pollution have been identified as including household and industrial wastes and waste materials discharged by tankers and ships berthing at the Ethiopian ports. Proper provision for the carrying out of studies related to sea pollution will need to be made in order to assure that appropriate rules and regulations are introduced in time to protect the much needed sea resources and assist in the development of tourism. In fact, at present, there are some pertinent regulations already in force in this area, but they require substantial revision.

In Ethiopia, several government ministries and agencies are involved in the various development aspects of the coastal area:

	Agency/Ministry		Tasks
(i)	National Water Resources Agency, Ministry of Mines, Energy and Water Resources.	(i)	Water resource studies and development.
(ii)	Ministry of Mines, Energy and Water Resources	(ii)	Geological exploration, and mineral and geothermal investigation.
(iii)	Mining Corporation, Ministry of National Resources.	(iii)	Mineral development.
(iv)	Marine Administration, Ministry of Transport and Communications	(iv)	Development of ports and regulation of related activities, e.g. discharge of waste and hazardous materials by tankers and ships.
(v)	Ministry of Public Health	(v)	Control of waste disposal into the sea from settlements and industries.
(vi)	Tourism and Hotel Investment Corporation, Ministry of National Resources.	(vi)	Development of tourism.

It is apparent that a number of government ministries and agencies have responsibility over the development of the coastal area. In the absence of proper co-ordination, it is likely that duplication of effort and waste of resources could occur. Also, in the absence of any comprehensive study on the economic use of the available resources such as land, water, minerals and fish, it is possible that development of these resources could be in conflict with each other and hence impede their optimal utilization. An integrated approach seems to be called for and some guidelines would be useful. For rational development planning, an inventory of all technical data and information should first be undertaken and analysed. On the basis of proper analysis, projects could be preliminarily identified and carefully studied for their comparative advantages and priority ranking.

viii. Finland

Information concerning the coastal areas of Finland is not separately available, because those areas have not been treated separately either in planning, the compilation of statistics or environmental protection. Population density and the degree of industrialization and economic development vary significantly in these areas. Large parts of the archipelago belong to the less developed areas of the country whereas some coastal areas in southern Finland belong to the most industrialized regions.

The need for planning applies to all regions of Finland and emphasis is laid on land use planning. The Division for Planning and Building in the Ministry of the Interior is authorized to co-ordinate land use planning on national, regional and local levels. Three detailed plans for recreation and environmental protection have already been prepared for coastal areas, and the rest of the country will be covered by similar plans by 1 July 1977. The density of vacation housing has been studied in order to define the need for land use plans on shore-lands.

The need to integrate the different economic and social functions in the coastal areas is considerable. It is particularly important to integrate planning for industry, environmental concerns and recreation functions. A special law for the archipelago has been proposed as a general measure of regulation. Some of the problems are due to accelerated industrialization and migration to cities which affect the economic balance of the area and bring about unemployment, thus creating the need for overall regulation.

Until recently, there have been no sweetwater basins in the coastal area, but some basins have now been built for the use of industry.

Soil elevation creates difficulties in the development of port facilities, especially on the Ostrobothnian coast.

Although pollution has until now been limited to areas near major rivers and industrial communities, vacation housing affects all coastal areas, thus damaging the quality of the environment and hampering land use for other recreation purposes.

The principal problems for port development are the shallowness of the approaches to ports and the difficulties created by ice conditions in wintertime.

Industrial siting puts increasing demands on the development of coastal areas. The most difficult problems are created by large industrial plants that require coastal siting, such as oil refineries, nuclear power plants and large shipyards. Their siting demands thorough investigation of other forms of land use. The regulation of industrial siting can be carried out primarily according to legislation on the protection of waters, regional planning and government subventions related to regional planning policies.

The State Committee for National Parks is investigating possibilities for creating a network of national parks in coastal areas, aiming at the preservation of characteristic natural features in these areas and in the archipelago.

The most essential problem for regional planning in the coastal areas is the high concentration of population, primarily on the coast of southern and southwestern Finland. Population density there is not very high in comparison with many other European countries, but the speed of migration has caused a number of problems. Due to a concentration of industry, there has been a shortage of labour in these areas whereas continuous unemployment has prevailed in other parts of the country. In the archipelago on the other hand, both employment opportunities and population have been on the decrease for several years.

For the purpose of applying various economic support measures to the less developed regions, these regions have been divided into two zones according to their degree of economic development. The law permits the formation of limited zones which can receive supplementary support. The island municipalities of southwestern Finland and the Aland Islands belong to such a supplementary support zone. In these areas increased governmental subsidies can be paid to support manufacturing, tourism, fur production, fisheries and fish breeding, horticulture, market gardening and peat production.

Laws concerning environmental protection especially in the coastal areas have not been passed so far. The general regulations included in the Planning and Building Act (1958), Water Act (1961), Public Health Act (1965) and the Protection of Nature Act (1971) form the legislative basis also for coastal development. Regulations on the protection of waters are furthermore included in the Act on the Prevention of Pollution of the Sea (1965) as well as in the Act on the Control of Oil Damage Caused by Ships (1972). These Acts may be amended when relevant international conventions, such as the Convention on the Protection of the Marine Environment of the Baltic Sea Area, have entered into force.

ix. German Democratic Republic

The German Democratic Republic's coastal area (i.e. the Rostock district) includes offshore islands, peninsulas, shallow bays and a relatively narrow coastal strip (17 to 60 km) along the Baltic Sea. The coastal line is composed of 341 km of outer coast and 1050 km of bay coast. In 1975, 870,000 inhabitants lived on a territory of 7100 km^2, making the coastal area one of the most densely populated economic regions (123 persons per km^2) in the north of the country. The coastal area comprises 6.5 per cent of the territory and 5.1 per cent of the total population.

Priority uses in the coastal area are shipbuilding, including ancillary industry; maritime traffic and harbour industry; fishery and fish processing; agriculture; and recreation and tourism. 3.2 per cent of the gross industrial production is produced in the coastal area, while 7.8 per cent of the arable land is situated in this area. Fallow land, which exists only on a small scale, is being reduced by recultivation measures. There exist only minor mineral resources (pyrites, sands and chalk).

In 1975, 2.5 million domestic and foreign tourists spent their holidays in the seaside resorts. This number will increase, as planned new capacity becomes operational.

The settlement structure in the coastal area is relatively balanced. Four cities with more than 50,000 inhabitants (Rostock, Wismar, Stralsund and Greifwald) and 38 townships and municipal communities with more than 2000 inhabitants form a regular pattern of human settlement over the territory of the coastal area. The aforementioned four big cities of the coastal area produce 70 per cent of the industrial output of the Rostock district.

Agricultural production was traditionally predominant in the coastal area. Since the late 1940's, limited shipbuilding, maritime traffic and harbour industries have been transformed into a rapidly growing sea-oriented industry including shipbuilding, maritime traffic, harbour industry, fisheries, fish processing and ancillary industry for shipbuilding. At the same time, agricultural production has been intensified. Recreation facilities have also been considerably expanded. This dynamic development implies rapidly growing requirements with regard to technological and social infrastructure, manpower and its specialized training. Development of industry, agriculture, recreation centres, towns and communities is taking place in the area as a whole and is relying on all natural and economic resources. For the area concerned, this does not necessarily mean more rational utilization; in fact multiple use could possibly result in conflicts.

Erosion is caused by wind, sea waves and littoral phenomena, short-term storm wave effects on biological and physical protective systems and ice pressure during long winter months. The necessary means are made available through government's financial assistance and the necessary capacities exist in the Rostock district. A system of protection is in use for example, where new and supplementary dykes have been installed for high-tide protection and and defence in areas endangered by overflow; artificial shore stabilization with sediments is being provided for endangered low coast areas. Scientific and technological measures are also used to control sea waves by using offshore structures in flat and steep coastal areas. Investigation has been undertaken to identify factors influencing the coast within the 5m depth line, the surf area and beaches, through testing mathematical models of natural processes. Adaptable and highly resistant construction material has been introduced to protect structures, and measures have been taken to improve high-tide and weather forecasts.

Port and harbour facilities at Wismar and Stralsund have been reconstructed and modernized, and a new transatlantic harbour constructed in Rostock. All three ports are used throughout the year and the navigable channel is frequently dredged. Both the Government and the Administrative Council of the Rostock district use all means available to minimize pollution of coastal waters and the Baltic Sea caused by waste waters from harbours, industrial plants and communities. These means include a system of waste-water purification plants, and certain specific research projects on, for example, the problems of vertical and horizontal mixture in waste product effluents resulting from currents and coastal reliefs and the installation of control systems by using computer simulation of aquatic ecological systems on the basis of light, temperature, nutritive substances and other parameters.

A complex planning approach exists for the whole of the Rostock district. For their management and planning activities, the competent authorities of the coastal area use data and information from scientific institutions (Institute of Oceanography, Board of Economics of Water Supply and Distribution, and others), central national economic plans, and regulations concerning general and special environmental protection.

x. Guatemala

The coastal area of the country has been treated as a separate entity with regard to zoning. The coastal areas extend 3000 m inland from the seashore, measured from the high-tide mark. The Pacific Coast is 254.7 km long and the coastal area covers 764.1 km^2 and has a continental shelf (up to 200 m in depth) of 14,700 km^2. The Atlantic coast is 148.1 km long (not including Belize), with a coastal area of 544.3 km^2 and a continental shelf of 2100 km^2. The total area amounts to 1308.4 km^2 with a total continental shelf of 16,800 km^2.

The priority uses of the coastal area include commercial activities; salt production; recreation and tourism; harbours; nature conservation; ecological preservation; and refineries.

An inventory of coastal resources including basic economic, social, technological and climatological data, does not exist. There is a need for a land register covering the leases granted by government to determine the land ownership system in the Pacific coastal areas covering a length of 254.7 km and an area of 764.1 km^2, and a continental shelf of 14,700 km^2.

The Pacific coastal area lacks so-called protected waters (gulfs, bays or inlets) and the whole year-round strong wave action has an impact on landing operations. From a geological and topographical point of view, the coastal area as a whole forms a plain, since the Pacific Ocean slope has only a very slight inclination.

Recently, a specific Office for the Control of National Reserves (OCREN) was established. 17/ Overall regulation is called for and it is understood that the functions of OCREN include the planning and regulating of national reserves.

The coastal areas are short of fresh water because the water found in tidelands or estuaries, bars, river mouths and canal and gully outlets is brackish. The coastal areas are also subject to the hazards of river flooding, torrential rains and heavy erosion caused by rivers flowing into the sea; this erosion is greater on the Pacific side.

xi. India

In India there are five levels of activity with which any coastal development scheme must be concerned: central government, the national corporation, the state, the municipalities or village councils, and the private owners or occupiers of the land.

Central government has full powers not only in union territories, but also by delegation and otherwise in the states (or provinces). Planning as such is both a central and a state concern, but the implementation of the plans may be the work of agencies set up in union territories and in the states by central government. In such diverse areas as archaeology and industries, central government may take the initiative regarding an area and exclude other and competing interests.

Central government's purposes may be achieved through the establishment of national corporations, some of an advisory character, some executive. To enable these to function, the special powers of government are available. For example, if an oil refinery is needed in a coastal area, government will acquire the land for it, or the corporation may do so itself if government's powers are delegated to it. The basic scheme common to India, Pakistan, Bangladesh, and in some measure Sri Lanka is as follows: once a field for activity is identified, a corporation (a Board or Trust) may be created by a statute. This body would be the channel through which the politically-conceived purpose is put into effect. The body is set up with its own constitution with such rights as to hold and dispose of property, to borrow funds and to lend, and to acquire property. The constitution and working and dissolution of the corporation are regulated by rules promulgated from time to time by the Government under which it is set up. The corporation operating within these statutory rules makes its own regulations or bye-laws for its operation, and for its dealings with the public. The public are governed by the regulations as if they were contained in the statute. The obligatory nature of both rules and regulations is exactly the same as that of the statute. The corporation is armed with powers to take positive actions and to prevent its functions being frustrated. The statute usually provides penalties for interference with the corporation's activities and protects the officers from suit for damages when they act in good faith.

State government has powers of an equally pervasive character provided it does not trespass on ground pre-empted by central government or in fields reserved for

17/ Decree 35-73 Act regulating national reserves.

central government in the constitution. The state likewise can create agencies or corporations in, for example, town planning, and housing, which exercise powers of government within their particular scope. The powers of the corporation are similar to those already mentioned above.

The municipalities, village councils and the like exercise wide powers with reference to amenities, and they may even play some role in planning. The fact that in practice these bodies are dominated by the state government does not alter their powers or responsibilities. Generally, all coastal areas come within the jurisdiction of a municipality or a village 'panchayat'. Questions such as open spaces, roads, sanitation, buildings, drainage, water supply or reclamation, all fall within the competence of local government. Any coastal area development scheme involving these matters must seek co-operation from, and indeed, implementation by local government.

The Housing Boards play an important role in land use and development. With the exception of the slum areas, the Housing Boards have, alone or in conjunction with other organs of government, powers to undertake housing schemes, to purchase and lease land, to evict occupiers and to enter for survey. 18/ Housing and improvement schemes may provide for recreation and transport facilities, sanitary arrangements, conservation and prevention of injury to or contamination of rivers and other sources or means of water supply, reclamation or reservation of land and parks, and fuel depots. Any coastal development project could presumably be undertaken within schemes by the Housing Board. A large number of tourist schemes have included rehousing, street construction, land development, co-ordination of housing activities, acquisition and disposal of land. Town planning is often a part of a comprehensive housing planning enactment. This does not mean that all aspects of town planning are subsumed. There is separate legislation 19/ under which the competent authority is given power to requisition and acquire land for any "development" purpose. Some local laws authorize a master plan to be set up or provide that in notified areas, no development of land may take place without application to the planning authority. 20/

In this connexion, reference is made to the Dacca Metropolitan Development Authority Ordinance of 1974. Although Dacca is not a coastal city, the Ordinance would appear to be instructive. The Dacca Metropolitan Development Authority (which deals with an ancient city haphazardly developed since 1947) prepares a master plan relating to land use, zoning, and land reservation; water supply, sewerage, and drainage; roads, highways, and traffic circulation; community planning, housing, slum clearance, and slum improvement. All land use and development must then be in conformity with this plan. All buildings and land are vested in the Development Authority. The Authority may take measures and exercise powers to carry out the purposes of the Ordinance. It co-ordinates schemes and supervises their execution. Thus, the Ordinance empowers the Authority in the widest possible terms, and clearly and unequivocally erects its supremacy in the field covered by the approved plan.

18/ See for example, West Bengal Housing Board Act, 32 of 1972, amended by 35 of 1973.

19/ See for example, (Kerala) Town Planning Laws (Amendment) ordinance, 1976; Bengal Development Act, 16 of 1935 as amended by 1 of 1963; the West Bengal Development Corporation (Amendment) Act, 20 of 1963.

20/ See for example, the Orissa Town Planning and Improvement Trust Act, 10 of 1957, amended by 23 of 1969 and the Bombay Town Planning Act, 27 of 1955.

The real difficulty for development in South Asia is not the resistance of landowners and other vested interests, but the conflicts between different local and national authorities, and perhaps the lack of clear mandates and the need for better co-ordination. A simple statute like the above strives to clear the ground and make a fresh start, at least in structural terms.

One of the basic questions is whether land could be acquired for coastal area development. The Land Acquisition Act is the basic statute. 21/ The basic purpose of the Act is to enable land to be acquired when it is needed for "public purposes" defined in general terms or for the use of a company specifically created by a statute. Where it appears to the appropriate government that land in any locality is likely to be needed for any "public purpose", a notification is published in the Official Gazette and the Collector ensure public notice is given in the locality. The Collector causes the land to be marked out and measured, and a plan to be made. He enquires into objections and awards compensation, and then takes possession. Those who do not accept the award may require the Collector to refer the matter to the Court.

Some Indian states have their own acquisition acts and their powers are sometimes more specific. For example in Kerala, land may be acquired for established projects, and state government also has the power to acquire land in project areas to avoid sudden rises in land values. 22/

Eviction of squatters and slum clearance can be used as a vehicle for development. 23/ Under the general scheme, a "competent authority" may declare any area to be a slum area, and require buildings unfit for human habitation to be improved. Central government, or the representative of the competent authority, may acquire land with compensation and make it available for development or other purposes. As a rule, the State Housing Board ceases to exercise powers when a slum improvement area has been declared and the Slum Improvement Board has begun to function in respect of it. Panchayats (village councils) may be established to carry out improvement functions. This machinery permits a degree of participation by the public. Government can always supersede a Board when it fails to function. However, it leaves the Board to determine which are slum areas, and how to deal with them. The Board is vested with the power to implement its decisions by positive action and negatively by preventing contrary action on the part of individuals. The availability of the compulsory purchase of land for slum clearance is a powerful means if any development project is intended.

In 1974, an Integrated Area Development Plan was prepared for the Pondicherry Territory. Special attention was paid to questions of fisheries, sanitation, and

---

21/ Land Acquisition Act, 1 of 1894. The Act was amended in India by 31 of 1962; in Pakistan, there are Land Acquisition Amendment Acts of 1957, 1965, 1968 and 1969. In Burma, the Act was amended by 67 of 1954.

22/ See two Kerala Land Acquisition amendments: Land Acquisition Act, 21 of 1962 and President's Act 4 of 1966.

23/ For example, (India) Public Premises (Eviction of Unauthorized Occupants) Act, 32 of 1958, the Amending Act, 32 of 1968. The term "public premises" include property of any company, 51 per cent of whose share are owned by the government. Slum Areas (Improvement and Clearance) Act, 96 of 1956, amended by 43 of 1964; Calcutta Slum Clearance and Rehabilitation of Slum Dwellers Act; West Bengal Act 20 of 1958; Madras Slum Improvement (Acquisition of Land) Act, 11 of 1954.

water supply. The result was that the Planning Department of the Government of Pondicherry is now in a position to put an end to casual and unplanned development of the town and port of Pondicherry and its environs.

xii. Indonesia

A large proportion of the Indonesian population lives in or adjacent to the coastal zones. Population pressures combined with demands for multiple uses of coastal resources - such as urbanization, industrialization, shipping harbours, power plants, recreation, fisheries and aquaculture and nature conservation - result in conflicts of interest. In addition, the impact of human society with its technological manipulations has almost always been accompanied by undesirable effects. However, since changes are inevitable, it is necessary to develop the best means to ensure that they will have the least possible impact, and that ecological principles are closely followed in their implementation.

The coastal zone of Indonesia is an interphase or ecotone environment. Consequently, many different ecological conditions can be found in a small area. This complexity creates management difficulties since generalities do not apply to all elements of the system, and capacity for different uses varies. Realizing the complexity and the urgency of the problems, and to generate interest in an integrated planning and management of the coastal zone, a series of symposia have been organized by the National Committee on the Environment in co-operation with research institutes and universities. It is fully realized that in order to make management programmes effective, a coastal management zone must be defined, and a set of management guidelines for various control activities has been formulated.

So far as the planning process is concerned, there is the National Planning Board at the international level, the Local Planning Board at the provisional level, and the Bureau of Planning at the ministerial/sectoral level.

The coastal zone is under the control of many agencies as well as individuals, who may act independently in their decision-making. There is limited co-ordination and sometimes overlapping in decision-making control. Even with the involement of many agencies there is no guarantee that all important resources in an area will be considered worthy of conservation. Nor is there any guarantee that net benefits will accrue to the rural coastal smallholders who are likely to be the group most sensitive to ecological changes. Strategies of coastal zone management must deal with the overall set of problems in the regulation of resource use and economic development in an area. The strategies must be specific, yet comprehensive, extending into the regulation of both upland and offshore root causes of coastal problems and bringing co-ordination to existing programmes in the coastal zone.

The following issues are being considered: Who should be responsible for devising and monitoring coastal management strategies? Is the authority to be split among the existing agencies having sectoral interests in the coastal zone? Should it lie with local-level planning authorities? Is there a need for a committee at the national level which can evaluate strategies, suggest standards and otherwise guide management? Should there be a full-time staff specifically concerned with coastal area development in some central government unit?

xiii. Jamaica

The relatively small size of the island imposes acute spatial constraints on its ability to increase production in the face of mounting population needs. Due to its rugged interior highlands the major impact of development exploitation has occurred in the limited coastal zones.

Jamaica has high rainfall and the resulting water is available from surface and underground sources. Although the supply is more than sufficient for the needs of Jamaica in the forseeable future, there exists the problem of its conservation, development and distribution, since the areas of greatest water demand have the least abundant supply.

At present, there is no serious problem of natural erosion of Jamaican beaches outside of normal periodic or random physical processes. However, the potential for future problems exists due to the destruction of reefs by sewage disposal in their vicinity, the construction of artificial harbours and beaches, physical encroachment on the foreshore and the illicit removal of beach sands. All these factors tend to alter the foreshore dynamics.

The expansion of the city of Kingston has brought about the filling of extensive areas of mangrove swamps and tideland areas, with a corresponding modification in the water basin structure. Major redevelopment of the Kingston waterfront has concurrently taken place, involving extensive dredging and land fill, and a causeway has been constructed across the entrance of Hunts Bay. The latter has affected the natural water circulation pattern between Hunts Bay and the harbour basin. Similar physical alterations to the waterfront are being carried out in Montego Bay and Oracabessa.

Tourism has become the island's second largest foreign-exchange earner, after bauxite and alumina. Long-term projections indicate the need for a total of 38,000 rooms and an employment potential of 65,000 by 1990. The full impact of this projection is yet to be studied. Excluding the social and physical problems associated with such development, an expansion of tourism without due consideration to the ecological implications, can have an adverse effect on coastal resources. Areas of mangrove swamp have been destroyed for the construction of hotels and resort facilities. The attendant sale of conch shells and coral has been referred to before, but further damage to the reefs is caused directly through collecting by tourists. Most, if not all hotels located on the shoreline, affect secondary treatment of sewage, and such treated effluents are discharged in a manner prescribed by the relevant government agencies.

Some of the major problems facing Jamaica relate to soil erosion, urban run-off, flooding, pollution from agricultural, domestic and industrial sources and the destruction of ecologically sensitive areas. These problems reflect poor land use practices, which in turn reflect the poor general understanding of the nature of land and the sensitivities of its resources. There is a need for an improved awareness of the country's natural resources and for the institution of a development control procedure, by which the normal functions of the physical planners would be strengthened by the addition of ecological skills.

It is important that there should be proper recognition of the ecological factors that comprise the island's economic systems, and of the limitations of its resources. This applies especially to the resources of the fragile coastal areas which suffer most heavily from the demands of population pressure and development. Jamaica's coastal resources include beaches, locations of scenic and cultural interest, black sands, estuaries, wetlands, marine grass beds, coral reefs and other marine organisms.

The plan for coastal management endorsed the evaluation of sensitivities and classification of areas of environmental concern. It also recommends the formation of guidelines for the allocation of coastal space among competing lines, and that the following considerations should be taken into account: (i) increasing the public awareness and understanding of the island's ecological systems; (ii) collecting, storing and distributing data and information on the development and conservation of the island's coastal resources; (iii) determining the policy guidelines to be followed and standards to be maintained in the management of the island's coastal resources of land, water, air, flora and fauna; (iv) promoting and ensuring the wise use of the nation's coastal resources by the establishment of an ecological review procedure for all relevant development proposals; (v) implementing programmes for the conservation and development of all coastal areas.

xiv. Mauritius

Little co-ordination exists at present between different government sectors concerned with the different aspects of coastal development, e.g. hotel development, industrial development, aquaculture, etc.

Cyclone and flood damage are major constraints to the development of aquaculture, due either to physical damage to the enclosure or to rapid salinity fluctuations which kill fish. Beach erosion is a problem requiring constant attention in regions with strong coastal currents; dredging should be regulated, and the construction of jetties, which affect the current patterns and cause erosion, requires careful study.

Marine pollution is not yet an acute problem in Mauritius, but stringent regulation of industrial development should be effected to prevent any increase in pollution. It has been found that even fresh water released in substantial quantities can affect coral growth and fish nursery areas.

Mauritius has a high population density and high unemployment. Although this condition does not cause any significant degradation of the coastal area from the fisheries point of view, it does result in a large number of subsistence fishermen entering the lagoon fisheries, with consequent overexploitation of fish stocks.

Biological primary productivity is low around Mauritius, and the result of this is that fish production is low and the possibility of aquaculture development limited. Conversely, the extremely clear waters form a major asset for the tourist industry.

The major problems unique to the coastal area are (i) the restricted area of good beach in relation to the total population of Mauritius, particularly in the light of a vigorous programme of tourism development; (ii) ribbon development along the coast as a result of lack of proper planning control in the past; (iii) until recently, uncontrolled subdivision of land for residential purposes within a strip of land immediately behind the coast; (iv) overfishing of lagoons; (v) damage to shell and bird life as a result of overcollecting and vandalism.

The coastal area of Mauritius is regarded for physical planning purposes as both a separate entity and a part of national planning. The coast is dealt with as an entity in respect to recreation. But for housing development, sanitation, water supply and education, the problems in the coastal area are regarded as the same as those in the rest of the country, and come under the Town and Country Planning Act of 1954. Coastal land is part of state property, and planning can take place through decisions on the leasing of land on the coast. For planning purposes, coastal areas begin from 1 km inland of the high-water mark to the end of the coral reef (about 50 m). The Town and Planning Ordinance is the principal law governing the reservation of land for roads, transport, communications and other public utility services. On the basis of this Ordinance, existing use pattern on the coast has been analysed and proposals for future uses are being made. There are separate laws governing protection of historical objects, building construction, sanitary conditions and removal of sand. These laws are applicable to the coast as well as the rest of the island. Special regulations also exist regarding port areas. For example, along Port Louis the coastal land within the city boundary is assigned for harbour and industrial uses only.

xv. New Zealand

General problems in the coastal area stem from the area being a fixed resource which is subject to an increasing range and intensity of demands. These demands include pressure from land-based activities in rural, industrial and urban development, roading, waste disposal and outdoor recreation, together with their consequent impact on adjacent water areas and on the water. They are a product of an increasing population, continued economic growth and social changes and consequent higher incomes and increasing leisure time.

The coast is highly valued for recreation and retirement purposes. With an increasing population together with changing demands, the basic problem is how to accommodate these together with other demands without destroying the natural resource. The impact of these demands is concentrated on particularly attractive areas, those with greatest ease of access and those near to population centres. Specific problems such as pollution and erosion are not widespread throughout the country but are concentrated in particular locations. The following problems are significant within the New Zealand context, although the magnitude of these is small when compared with some other countries.

One of the most critical problems has been the pressure on coastal resources from residential development, particularly for holiday purposes. In many cases too much coastal land has been committed ahead of genuine needs for urban development without consideration of real needs. The number of sections subdivided is greater than the number being actually used for building and this has led to haphazard development which is scattered and untidy and which leads to high costs of servicing.

In many localities single-unit residential development has been permitted at most beaches which are accessible by vehicular traffic. This reduces the variety of experience in the coastal area that people can enjoy and it only caters for a small proportion of demands from a growing holiday population. Often, little attention has been paid to siting buildings in relation to the surroundings, frequently destroying visual amenities. Beach development has also been permitted in areas which are susceptible to erosion, or they have been permitted with little thought to providing adequate sewerage facilities or to suitable public access for public recreation. The pressure to subdivide land for residential development and the consequent inflation in land values (see paragraph 4) leads to high costs and makes purchasing desirable land for public reserves more difficult.

Better planning is required for coastal areas. In respect of water, for example, there is a lack of overall co-ordination for controlling uses of coastal waters and the seabed, which is spread over a range of government departments, harbour boards, regional water boards, catchment authorities and territorial local authorities. Further, there is a need to ensure that this is recognized and taken into account when planning for coastal areas because of the interrelationship between land and water uses. Control of coastal land is primarily the responsibility of territorial local authorities. Their performance in planning varies considerably. In some key areas considerable progress has been made in adopting and implementing objectives and policies which recognize the importance of the coastal areas, with the introduction of techniques such as appropriate zoning, reserve requirements, and setback lines.

However, there are still many issues that point to the need for better planning. There is a need to incorporate objectives and policies which recognize the importance of coastal resources, within both the local and the regional contexts and to implement suitable planning controls to meet and direct varying demands. Often the administration tends only to recognize 'local' and immediate interests, to the detriment of wider considerations and the national interest. With demands often being generated from outside local authority boundaries, there is a need to plan the development of a resource from a regional point of view within the context of a national framework. However, there is often little or no co-ordination between local authorities responsible for coastal planning. Further, regional planning is weak in New Zealand because of inadequate provision in the Town and Country Planning Act, and because of the fragmented nature of local government. Also, the present planning legislation is oriented towards district planning and makes little reference to planning at the national level.

Another problem is the inadequate provision of qualified staff to advise on planning issues and the possible means of contending with them. In some areas, technical support is good while in others, weak or spasmodic. In general there is a shortage of qualified and experienced staff covering a range of technical fields at all levels of government.

Implementing better planning for land and water use alone will not necessarily contribute to better resource use. The general question of finance is a problem which pervades many of the issues in coastal planning and is manifested in a number of ways and unless this is solved, it would be unrealistic to expect that 'better planning' will contribute to solving problems. There are financial pressures on local authorities and landowners, which tend to force development. For example, due to inadequate income, local authorities look to residential subdivision as a means of increasing revenue, and they become reluctant to adopt policies which direct and/or restrict this process.

By way of second homes, hotels, motels, camps and camping sites and their associated uses, an unsatisfied market for recreation accommodation increases the value of land to a level deemed by the market to be suitable for these purposes, but one which is considerably above that pertaining to its traditional use. This places pressure on local authorities to provide these uses in accordance with the expressed demand and to approve undesirable subdivisions. Landowners can find existing uses of their land increasingly uneconomic relative to the enhanced land value. This is because the rates are too high in relation to the income derived from the land. Rates provide the major source of local body finance and the rates structure is related to the valuation of the land. The opportunity for landowners to realize profits on their assets by sale or subdivision becomes increasingly difficult to resist.

This situation gives rise to a high level of speculative activity and contributes to making any acquisition of land for reserves unrealistically expensive; a lack of adequate finance at all levels of government for the acquisition and development of reserves exacerbates the difficulty of purchasing land for public recreation. Central government has provided a fund of $1 m for five years, although with increasing land values this is completely inadequate.

In many areas where beaches for recreation represent a prime element in the panoply of coastal resources, the only restraint on a subdivisional pressures is inaccessibility. The improving or upgrading of road access has put pressure on areas to be developed. There is a need to ensure that land use in coastal areas and the degree and type of access (roading, walkway, water) are planned together. Too frequently, this factor has not been recognized in many of the important areas, where all the beaches with road access show some degree of adverse development.

Ports and harbours are foci for a number of particular problems and demands which all modify the harbour in some way. There are demands for reclamation to provide port facilities and marinas, conflicts between recreation uses and commercial development, and between different recreation uses. Water-related industries and waste disposal are in many cases located in estuaries and the importance of these sensitive areas is only now being recognized. Often, these water conflicts are accentuated both by the lack of planning of harbours as a whole and the failure to consider all the factors involved and provide a framework within which different demands can be assessed. Many of these demands are only considered on an ad hoc basis and not within the context of an overall plan for a particular harbour. Reclamation, in order to accommodate container facilities, needs to be assessed within the context of requirements for port facilities around the country as a whole and of likely national trade and economic requirements. Some of these issues are compounded by local authorities not taking a positive role in the planning of ports and in integrating their use with other planning policies. Some of the shortcomings in the present system are referred to below.

In general, there is a considerable amount of information available on many facets of coastal planning covering land-based activities, although it is dispersed over a wide range of agency and other sources, and needs to be drawn together and collated. Also, the data that are available may not be directly applicable to coastal areas.

A recent study covering coastal management concluded that there has been a serious lack of knowledge of the nature, quality and quantity of resources available to New Zealand in its coastal waters and seabed. For example, little is known of the importance and specific functioning of the life cycle of sea vegetation and animals in various sectors of coastal water areas. What research has been done has been applied to specific problems and limited areas.

Information is available on a range of activities affecting foreshores and coastal waters, and these include records of ownership and control, grants for all approved structures for foreshore and seabed, reclamation and licenses for removal of material from these areas (sand and gravel); also, data on tides and information on marine pollution are available.

There is also a considerable amount of data available on land use at all levels of government, though this does not apply specifically to coastal areas. Land use information is contained in local or regional schemes (see section B1). There is also wide range of information provided by government departments, including statistical data on topics such as demography and social characteristics, transport and communication, product-on and finance. Further, there is a considerable amount of physical land-resource information available in the form of land-resource inventories, from which land-use capability assessments can be made.

Figure 1 illustrates the complicated legislative administration of the coastal area in New Zealand, and shows the various laws and their interrelationship that come into play from the continental shelf to the hinterland.

The major tasks of the principal government agencies in New Zealand involved in coastal area activities (excluding fishing, and oil and gas exploitation) may be summarized as follows:

The Ministry of Transport (marine division) is responsible for approving plans of all work in, on, over, through or across tidal lands or the bed of the sea, or the bed or bottom of any harbour, navigable lake or navigable river. It is also responsible for the control of harbours, navigation and marine pollution. These provisions exist under the Harbours Act of 1950.

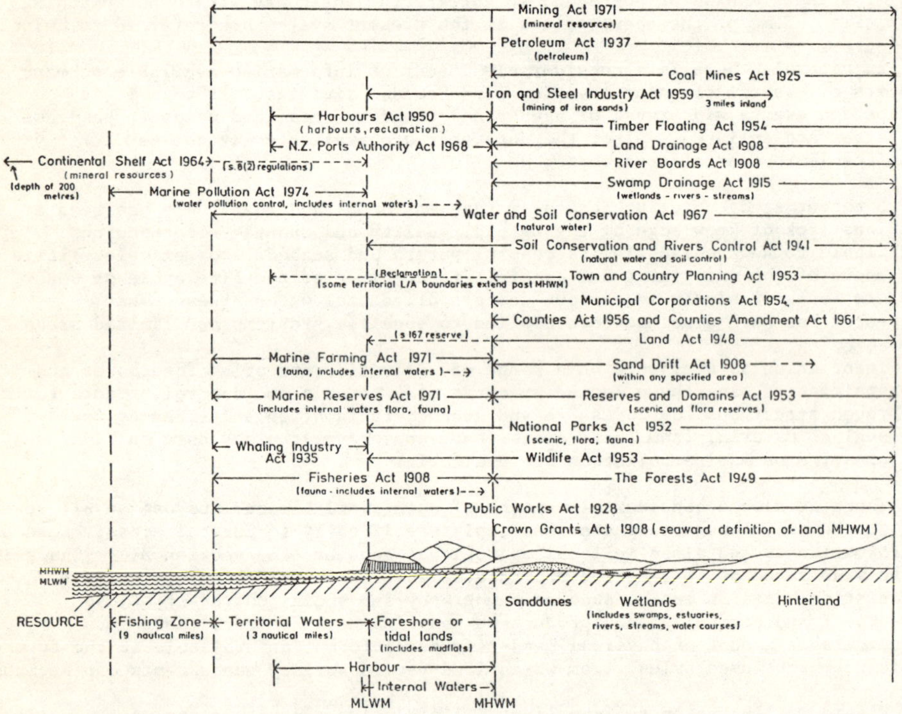

LEGISLATIVE ADMINISTRATION OF THE COASTAL ENVIRONMENT

Fig. 1

The Department of Lands and Survey is responsible for the Administration of Crown lands and different types of reserves - scenic, historic, flora and fauna. This includes national parks and the administration and management in maritime parks formed by the linking of seashore and island reserves. The Department is also responsible for carrying out on a national basis a coastal reserves survey. This identifies areas which should be protected because of their inherent qualities. It can make recommendations for administrative action, either to ensure the protection of the land or to bring it into public use.

The Ministry of Foreign Affairs in relation to coastal planning, is responsible for the administration of the Continental Shelf Act 1964 and the Territorial Sea and Fishing Zone Act 1965.

The Ministry of Works and Development is responsible for the administration of (1) the Town and Country Planning Act providing for comprehensive land use planning and control; and (2) the Water and Soil Conservation Act and Soil Conservation and Rivers Control Act.

The Commission for the Environment is responsible for the administration of environmental impact-reporting policy applicable to all government works, local authorities and private works requiring government finance or approval.

The Nature Conservation Council and Environmental Council are government-appointed independent advisory bodies, free to make recommendations to the government on environmental issues and to publish their views.

The New Zealand Ports Authority is responsible for fostering an efficient and integrated port and harbour system in New Zealand and planning for its development.

Local territorial authorities are responsible at the local level for land use, planning and control.

Harbour boards are responsible for the administration and management of harbours.

Catchment authorities (regional water boards, catchment boards and commissions) are responsible at the regional level for administering the Water and Soil Conservation and Rivers Control Act.

xvi. Nigeria

Nigeria's coastal area is endowed with a number of natural resources including crude oil, gas, trees (such as red mangrove and coconut) and fish.

The position occupied by the petroleum industry in the Nigerian economy is indicative of the importance of crude oil resources, and the ever-increasing demand for fish and fish products clearly signifies the need for greater fish and shell fish exploitation to boost production.

Nigeria's fisheries potential is found to be low, since the coastal area is for obvious oceanographic reasons not very rich in resources.

The sea along Nigeria's coastline is endowed with a number of natural resources. The exploitation of these resources contributes to the vast number of human settlements all along the entire 800 km coastline. Little information can be provided on the actual number of coastal settlements and population. However, based on available fragmentary data and extrapolation, it is assumed that there is a 2.5 per cent per annum growth rate in the population. The total human population along the Nigerian coastline is about 1,500,000 out of which 969,000 or approximately 65 per cent are fishermen; most of the remaining 35 per cent engage in farming and exploitation of coastal trees, including coconut and mangrove for food production and canoe construction.

The need for increased domestic fish production in Nigeria has resulted in concerted efforts to implement various fisheries development projects, aimed at aiding artisanal fishermen to increase their catch, to improve their boats and working conditions, and to adopt modern fishing gear and techniques. The problem of technology transfer in artisanal fisheries is also recognized. In programming the improvement of activities in this sector, caution is being taken to proceed at reasonable steps. The target for an increased coastal fish production has brought to attention the need for improved fisheries activities, particularly in the curtailment of spoilage, poor processing methods and inefficient handling procedures. The aim is to optimize the utilization of Nigeria's coastal fish stocks produced by artisanal fishermen.

In order to prevent oil pollution, regulations have been introduced to prohibit discharge of oil into Nigerian waters, and to install equipment to prevent oil pollution. Facilities at harbours can also provide for the disposal of all residue.

The collection of statistical information on scattered settlements along the coast is being organized by the Federal Department of Fisheries in conjunction with the Federal Office of Statistics.

To ensure the proper development of available shoreline, interministerial co-ordination exists in the planning of coastal area development and usage.

xvii. Panama

The most critical problems for the coastal area are (i) the lack of an information base necessary for planning; (ii) in some coastal areas, especially in the Bay of Panama, there have been flooding problems as a result of torrential rains; and (iii) pollution in the port areas of Mensabe and Aguadulce, especially in the Bay of Panama area, whose waters are polluted by industrial wastes and sewage.

The coastal areas of Panama are not considered to be a distinct region but parts of each existing regional division. There is a preliminary draft Sanitary Code to regulate the use of coastal areas, but there are no specific regulations covering beach access, which is under municipal control; sand and gravel extraction is under municipal control. Traffic of ships carrying hazardous materials has been ratified by Panama but there is no specific national legislation on the subject.

The priority uses in the area are commercial activities, recreation and tourism, harbours and waste disposal. No regulations have been established concerning industrial site locations. There is a proposed policy on industrial site location which would prohibit the location of highly-polluting industries in the coastal area.

The Renewal Natural Resources Directorate (RENARE) of the Ministry of Agricultural Development has laid down guidelines pertaining to planning with regard to beaches and dunes. The Ministry of Commerce and Industry and RENARE have prescribed protective measures for spawning grounds.

xviii. Philippines

The first critical problem is the lack of sufficient understanding of coastal processes and behaviour, such as estuarine circulation, erosion and deposition, effect of waves and surges, ecological interaction of living organisms and environment and other problems. This lack of adequate understanding results in the inadequate planning of coastal area development. There is also a need for effective governmental regulation in the uses and management of the coastal zone.

Although there are at present no serious problems of a freshwater shortage, the country may encounter this problem in the future. A Water Resources Council has recently been created for the express purpose of ensuring adequate supply of water in the years to come. The problems of flooding, erosion and seasonal storm damage are particularly of great concern. Annual losses to the economy in millions of pesos are caused by these natural hazards. Marine pollution is also becoming a critical problem in coastal areas principally because of landbased sources, mining activities, and industrial development, and is aggravated by the high population densities found mostly in coastal areas. All these problems contribute to the need for an integrated approach to coastal area management and development.

Regulations on the use of marshes and shoreland areas are issued by the Bureau of Fisheries and Aquatic Resources especially in the establishment of fishpond areas. Zoning laws exist mostly in chartered cities and populated districts, but there are no regulations on beach access. The National Pollution Control Commission is required by law to be consulted in the siting of power plants and in the establishment of industrial factories and complexes. The Bureau of Public Works is entrusted with the dredging of coastal areas especially in ports and harbours, while the Bureau of Mines issues and enforces regulations on the extraction of sand and gravel for building purposes.

The National Pollution Control Commission is entrusted with environmental protection in coastal areas, while the Philippine Coast Guard, in a recent Presidential Decree, has been given the task of preventing and monitoring oil pollution from vessels.

xix. Singapore

In Singapore the interaction between the sea which surrounds the country and the interior of the country is recognized to be very intimate. Therefore, coastal area planning has always been an important cornerstone of national planning. Increasing pressure on meagre land resources has forced national planners to look to the coastal area to obtain new areas by massive reclamation. The reconciliation of the varied and sometime conflicting requirements of important coastal activities such as port development, recreation, fisheries and the preservation of the ecological balance is thus one of the challenges faced by government.

The important areas of development activity in the coastal zones of Singapore are:

(i) Port Development: Singapore is the fourth busiest port in the world with a throughput of 61,268,900 and 60,420,300 tons of cargo in 1973 and 1974 respectively. Port development has already covered the bulk of the southwest and southeast coastline of the island as well as part of its northeast coastline. Future major port development is expected to be located at the north end of the eastern coastline. Some offshore islands, particularly those on which major oil refineries are located, also have their own ship terminals. In addition, offshore terminals have been created to cater for the very large tankers.

(ii) Land Reclamation: Major reclamation has been carried out, is in progress or is planned for the development not only of port works but also of such other activities as airports, major coastal highways, industrial sites, large housing colonies, parks, marinas and other recreational facilities. Reclamation has not been confined to the main island only, but has also been carried out around offshore islands and reefs.

(iii) Recreation: Inland recreational areas are rapidly becoming inadequate due to increasing demand generated by not only the rising population but also greater affluence. Due to other priority users, some of the coastal recreational areas have also become unavailable to the public. The Government is therefore developing offshore islands for this purpose and creating the communication facilities to enable the public to reach them quickly and cheaply.

Oil pollution is the most serious problem. There are five major oil refineries in Singapore, three of which are located on offshore islands and the rest in the foreshore areas. All of them have tanker berths in close proximity and from two of them, there are potential sources of oil pollution. However, strict control and constant vigilance have so far prevented any hazard. Other sources of oil pollution are the hundreds of ships in the port, especially when bunkering, and also some 35,000 ships which pass through the Singapore Straits which are contiguous to the Singapore port limits. However, the greatest threat of oil pollution arises from accidents - collisions and groundings involving loaded tankers/VLCCs in the Malacca-Singapore Straits as well as within port limits. In order to reduce, if not totally elimiate, pollution arising from such accidents, it is therefore imperative that a traffic separation scheme and other safety measures should be drawn up for that narrow stretch of water. In co-operation with neighbouring countries, these navigational safety requirements are being considered.

Risks of other types of pollution from the disposal of industrial and human wastes have so far not been serious, but may become so in the years to come. Various means to deal with such risks are under study.

In order to combat pollution and attendant damage to the marine ecology, continuous monitoring of atmospheric and oceanographic quality is necessary. Such monitoring presents certain unresolved practical and theoretical problems, such as laying down realistic standards for acceptable levels of water quality, the behaviour of oil on water, the effect on oil of natural processes such as evaporation, solution, oxidation and biodegradation, the spreading of oil in a still sea, the effect of winds, currents and waves, and the deployment of a suitable apparatus for measurement and monitoring, etc. This calls for increasing the information base by means of observations, experiments and research on local conditions.

The coastal areas of Singapore cannot be treated as a separate planning entity; rather, they constitute a very important and integral part of the totality of national planning. This is because Singapore is an island of small dimensions, and its coastal areas are therefore critical. If the coastal areas are taken to mean a strip two miles wide along the foreshore (excluding all rivers), this strip occupies more than 50 per cent of the main island and all the offshore islands.

Various laws are applicable to the coastal activities as well as to the state as a whole. The Planning Act and the State Land Act (sand and gravel extraction), for example, apply to the whole country, including the coast; the Foreshore Act and the Prevention of Pollution of the Sea Act are primarily designed to regulate coastal activities. Because of the size of Singapore, even the interior of the country interacts intensively with the sea. There is therefore a great pressure on the land and marine resources. The reconciliation of the varied and sometimes conflicting requirements of important coastal activities such as port development, recreation, fisheries and the preservation of the ecological balance appears to be one of the challenges faced by government.

xx. <u>Spain</u>

There is a need for advanced indicative planning, covering the evaluation of both natural and artificial resources and their tentative allocation to uses for which they seem most suitable.

Erosion and regression of the coast are taking place, caused by coastal dynamics or by human agency, and are resulting in the decay or even the disappearance of beaches, which constitute one of the most valuable of coastal resources.

Pollution is a serious problem and a notorious phenomenon which must be dealt with by means of large-scale co-ordinated planning of the health infrastructure and industrial purification.

High density of population on the coast is a problem and causes 'sprawl' and downgrading of the hinterland.

An integrated approach is certainly needed; however, in view of the multidisciplinary and complex nature of the coast, it would not be advisable to establish a single body or agency in charge of coastal matters; rather it would be better to deal with the problem by establishing overall, and specific co-operation for certain projects between various agencies, whose fields of competence should be very carefully delineated and which should include fully-operational technical and ecological units.

xxi. <u>Sri Lanka</u>

Several square miles of the country's coast have been lost through erosion, especially in the thickly-populated southwest coast, and this destructive phenomenon has now become a threat to life, property and communications. Coastal erosion in Sri Lanka has been aggravated by sand and coral mining and uncontrolled coastal activities (building drainage etc.). It is obviously futile to permit such operations to continue only to spend large sums of money to protect the coast. While existing legislation for sea-shore protection is weak, a solution must be found for those engaged in the illegal mining of sand and coral for a livelihood.

The causes of coastal erosion in Sri Lanka can be classified into the three categories of natural erosion, man-made erosion and biological erosion. Part of the natural erosion is of geological and long-range origin, and is associated with a tilting of the island and a rise of the mean sea-level. It has affected both the east coast and the west coast of the island, producing a predominantly headland-type coastal configuration which is dominant on the southwest coast and along part of the east coast. Another aspect of the natural erosion is the development of shorelines adjacent to tidal inlets. The most serious cause of erosion is man-made; it is related to the construction of outlets for drainage purposes, and most important, to the removal of sand from river mouths and the sea-shore, depleting the natural balance of sand of the coastal regime. The mining of coral along the southwest coast is particularly serious because it not only takes away materials which could become available as beach norishment, but it also destroys the coral reef that breaks down the high waves before they reach the shoreline. Another man-made cause that indirectly affects the stability of the coastline is the pollution of coastal waters by human and other waste. This degrades water quality and adversely affects (or kills) the growth of natural reefs. These reefs are present along the southwest coast especially in the Hikkaduwa area, where eroding tendencies are high, and along the middle part of the east coast, between Trincomalee and Batticaloa. Coastal areas subject to erosion come under coastal defences established by the government on the basis of priorities formulated within the existing policy of coastal protection.

Biological erosion occurs from coral-eating starfish which are found in abundance in the Trincomalee area. A programme to eliminate this danger to the coast is presently under execution.

The development of the tourist industry in Sri Lanka in the last decade has resulted in the demand for more beach recreational facilities especially from the many beach-oriented tourist hotels springing up along the coast, between Negombo and Tangalle on the west and between Batticaloa and Trincomalee on the east. Thus, in addition to the need to develop new beaches and to preserve existing beaches, the necessity to evolve measures to combat beach pollution, at least in the areas of tourist development, cannot be overemphasized.

The problem of conflicting interests in the coastal zone requires a broad over all view of the situation. The Coastal Engineering Research Centre of Sri Lanka has been proposed to co-ordinate research activities and to provide research services to all agencies that request them. Action has been taken to introduce new legislation for conservation of the coastal zone and a new division under the Ministry of Shipping Aviation and Tourism, responsible for the management of the coastal zone, is being organized for the planning and development of the coastal area and the effective implementation of the proposed legislation. The preservation, restoration and environmental control of the beaches and above all, the effective management of the coastal zone, has high priority in the country's development plans.

Priority uses in the coastal area can be classified as harbours, recreation and tourism, shore protection and waste disposal. Mineral resource development (ilmenite) on the northeast coast, some commercial activities, residential development mainly on the southwest coast and the north coast, and historical and archaeological sites in some of the towns are the next important uses in the coastal area.

Mineral exploitation in the coastal area is at present confined to the mining of ilmenite, sand and coral. Large amounts of ilmenite are present along a stretch of coast approximately 8 km long at Pulmodai on the northeast of the island. The sparce population in this area and the undeveloped character of the coast facilitated the legal mining of this mineral on a commercial basis by a goverment corporation. This activity, which commenced over a decade ago, is today a flourishing foreign-exchange earner for the country, and it has been estimated that adequate reserves are available for the next quarter of a century. With respect to the mining of sand, the present planning guidelines confine such mining to river mouths and lagoon entrances; it is carried out on the basis of permits issued by Government District Authorities. As regards coral mining, existing laws prohibit such activity within the foreshore and the bed of the sea in specified areas.

The fishing industry is located at several points around the island, and the planning for fishing operations has been based on harbour and shore facilities, fishing grounds in the vicinity, and traditional areas of fishing. In the development of the coastal area, preference has been given by government to fishermen for the use of the foreshore in certain traditional areas for net fishing and beaching boats, and the erection of temporary shelters. A few other industries are located in the vicinity of the harbours of Colombo, Galle, Trincomalee and Kankasenturai, to accommodate their dependence on raw material imports and to facilitate export of the finished products.

The tourist industry developed in the last decade is located at several points along the coast. Resort areas managed and controlled by the Ceylon Tourist Board have been set up at Negombo and Bentota on the west coast and Kalkudah on the east coast. Hotel development in these areas conforms to requirements set out by the Tourist Board.

Landfill and land reclamation in the country are only carried out at harbour development sites by state organizations. No offshore sand and gravel extraction is at present being carried out round the island's coast, while dredging in the coastal area is confined to the dredging in the coastal area is confined to the dredged spoil is sometimes deposited a few miles offshore at selected sites.

Ports of significance in the country are Colombo, Galle, Trincomalee and Kankesanturai (in the north). The development of these ports is undertaken on the basis of priorities determined by economic needs. No marinas exist at present in the country.

Excavation or removal of material from the foreshore or seabed is prohibited in specified areas by existing laws. The effective control and management of the coastal zone, which includes beaches and dunes, coastal reefs, scenic vistas, recreation areas etc., would form part of the responsibilities of the new Coast Conservation Division. The zoning of the coastal area would also be one of its functions.

xxii. <u>Tanzania</u>

Coastal towns usually dump their raw sewage into the sea for economic reasons. The large volume of water assists in diluting the wastes and the salt content of the sea acts as a natural sterilizing compound. For moderately-populated towns, dumping may not have a deleterious effect on the marine life, but with growing populations, action has to be taken to safeguard marine life while continuing to attract bathers on the beaches. The types of wastes associated with coastal areas are:

(i) the human environment sewage disposal;

(ii) industrial wastes - toxic and other solid wastes dumped into the sea from different installations;

(iii) port development - operations of cargo during loading and unloading can pollute the harbour area and ultimately the port surroundings;

(iv) oil spills and the dumping of chemicals at sea pollute coastal areas through wave and tidal action.

While pollution may not be a great threat at the moment, emphasis should be placed on making a proper study of the environment around the coast. The launching of the United Nations Environment Programme (UNEP) and the establishment of the FAO/SIDA proposed Regional Laboratory on Aquatic Pollution Control show the great emphasis being placed on controlling the dangers expected out of pollution. The East African Marine Fisheries Research Organization (EAMFRO) has been collecting data on oil pollution on the coast as a basis for preserving the environment. The studies to be undertaken by any organization should be directed to solving problems connected with rapid urbanization on the coast, anticipated port development, industrialization and the preservation of marine ecology and fishing industries.

The long coastline has many important rivers which create ideal conditions for the development of estuarine aquaculture, especially for highly valued species like prawns, lobsters and oysters. While the ocean resources with pelagic and demersal remain underexploited, mariculture can be initiated to utilized the vast areas of swamps which are at present largely unproductive. There is no study that has covered the role of aquaculture development in the fisheries of the country.

In 1967 an idea to set up marine national parks to protect and conserve marine flora and fauna was initiated. With assistance from United States based foundations and the United Nations Educational, Scientific and Cultural Organization (UNESCO), expert surveys have been undertaken to identify suitable areas for marine parks (marine reserves). Initial areas identified for marine reserves are the reefs around the islands lying close to Dar es Salaam and the coastal tourist hotels, and the area round Mafia islands and Maziwi Island. The Fisheries Act of 1970 emphasizes the need to establish marine reserves and the need to develop marine parks in a co-operative effort between the Fisheries Division and the National Parks. The Fisheries Division would be responsible for the management and conservation of the whole coastline while the National Parks would take measures to protect and develop specified tourist areas. The Fisheries Act also stresses the use of proper fishing methods, enforces a total ban on spear gun fishing and the use of destructive fishing methods (dynamite) and bans on the collection of live shells and corals.

Fishermen have been organized all along the coast into Ujamaa fishing villages and government assists by providing relevant facilities like fish-receiving facilities (cold room, ice-making machines etc.), training in the use of improved gear and craft and organized bookkeeping. The villages are able to make good use of experts attached to the village by the Fisheries Division. Throughout these villages, electricity, portable water and schools have been introduced, as observed at Moa Ujamaa fishing village. Initially, the villages also received a financial subsidy from government through the Regional Development Fund (RDF). Well-established development corporations can also be observed all along the coast (Pangani Development Corporation, Dar es Salaam Development Corporation, Mtwara Development Corporation etc.). The corporations utilize modern fishing gear and craft and are catching substantial quantities of fish and prawns. The corporations are financed through the Regional Development Fund and have access to loans from the Tanzania Rural Development Bank. Besides fishing, the development corporations may be involved in other commercial activities, such as the running of butcheries.

Any programme to develop and manage the coastal resources of Tanzania has to take into account the topography of the continental shelf which makes it difficult to exploit resources, the existing infrastructure and the government policy of socialism for self-reliance.

There is a need to carry out full-scale resource investigation to collect data on the shore and reef resources. Present investigations by EAMFRO, the Fisheries Division and the University of Dar es Salaam emphasize the improvement of fishing through better design and use of good-quality material; the modernization of local fishing gear and craft; fishermen-training to give fishermen necessary knowledge on the use of modern fishing equipment and the planning of improved port terminal facilities for the processing, storage and distribution of the catch, but the response has been slow. There is also an urgent need to study the social and economic problems of the fishing community so that these may be considered with economic planning.

The existing institutions should be encouraged to co-operate in collecting research data and in utilizing the available manpower and equipment. The fishing community should be made to take an active part in utilizing research findings to ensure optimum output from the scarce funds being invested.

xxiii. Thailand

Tin mining in the coastal areas, particularly in Ranong, Phanga and Phuket, has become an increasing problem, as it conflicts with the mangrove industry. After an area has been mined, the land generally becomes useless, and it takes a considerably long time before the land is turned to original forest. The tin mining operations on the bottom of the coastal sea also present a serious problem by causing deterioration of the benthic environment. Measures to control such operations are being discussed.

At present, there is no single government agency which is responsible for coastal area development and management. Each 'resource' is under the authority of the corresponding department. For example, the fishery resources are protected and managed by the Department of Fisheries while the Royal Forest Department takes care of the mangrove areas. Ports and shore facilities are under the jurisdiction of the Harbour and Marine Department. It is only when a conflict of interests occurs that interdepartmental co-operation comes into action.

With the growing use of sea lanes for commerce and the use of coastal areas for industrial development, the threat of pollution and other deterioration of the coastal areas become more acute. Various tanker disasters, such as the Santa Barbara oil well seepage and other accidents, have demonstrated the harmful effects that can arise in coastal waters and on beaches. Utilization of coastal land for industrial or community development has led to a deterioration of the ecosystems and has adversely affected the yields from the fisheries and aquaculture projects. However, knowledge on the various coastal ecosystems is still inadequate at present. Attempts have been made by the National Economic and Social Development Board (NESDB) of Thailand in 1975 to identify environmental criteria and policies for coastal zone management and to develop a detailed guideline for regulating new development in the coastal zone of Thailand. As a result, a proposal was made in which three major coastal management zones, those of preservation, conservation and development, were being considered.

Another proposal was made recently by the Harbour and Marine Department of Thailand on the establishment of a central pollution control center to co-ordinate and direct government agencies, government enterprises and private sectors concerned in handling pollution cases, particularly of oil spill. This proposal however is still in the drafting stage although it is very likely that it will be accepted by the National Environment Board in the near future.

xxiv. Togo

Better planning is required in order to remedy the result of unplanned development on the coast. An overall management plan of the coastal area is envisaged in the third five-year plan. All aspects of development will be taken into account, including, the interstate highway linking the coastal states, the erosion of the road between Lomé-Aneho, problems resulting from oil refineries, tourism facilities, and the preservation of forests.

In the control areas of Lomé and Aneho, there is a shortage of fresh water partly because of the density of the population and partly because of limited supply - only the top layer of the water is unsalted.

Flooding is quite frequent in the coastal areas, particularly during the rainy months when the rivers, lakes and lagoons overflow. The lagoons near Lomé and Aneho are being treated, but are not completely immune from the danger of flooding. There is serious shoreline erosion near Lomé, Baguida and Agbodrafo.

Marine pollution is still moderate but will worsen as new oil refineries are constructed.

An industrial zone of 600 ha has been established for togonese industries and for the use of the neighbouring states of Niger and Upper Volta.

The coastal areas are difficult to gain access to from the hinterland because of the existence of lakes and lagoons. A better transport system is needed.

PART II

SECTION TWO

Survey of Legislative Measures

CHAPTER 10

Adaptation of Town and Country Planning Techniques to the Coastal Area

In view of the rich experiences in town and country planning, many countries have adapted the techniques in this field to their coastal areas. As will be seen, 1/ this approach is similar to the United States' Great Lakes approach. The techniques used include the establishment of building or non-building zones; the designation of protected zones in which planning schemes are subject to particular regulations; the imposition of building prohibitions; compulsory acquisition of land as methods for controlling urban development in coastal areas; the incorporation of the communication network as part of the general planning; the designation of locations for industrial complexes; and the establishment of special regulations for tourism, e.g. public access, camping etc. 2/

Certain regions of Italy have promulgated, for example, laws which prohibit any construction on the coast measured 300 m landwards from the point of the waves. 3/

In Norway, central government is responsible for the preparation of national guidelines for the use and management of the country's natural resources and for establishing criteria for settlement patterns. These national guidelines provide the basis for county, regional municipal, and local planning. A local plan is a detailed zoning plan for a given area and covers all aspects of the physical environment within that area. The plan defines and delimits residential, recreational, industrial and commercial areas and parks. No development may occur when it is in conflict with a local plan.

1/ See Chapter 14.

2/ See J.A. Steers, Protection des zones côtiéres, Le Centre Européen d'Information pour le Conservation de la Nature, 1974, 87 pp. Amselek et Cohen, Mesures Législatives Prises ou à Prendre par Les Etats-membres du Conseil de L'Europe pour la Protection des Côtes, ibid., 1974, 149 pp.

3/ Two specific studies may be cited here: F. Fiorelli, Etudes italiennes pour le projet concernant les bandes côtières méditerranéennes, Instituto di studi per la Programmazione Economics; Les phénomènes d'utilisation et de dégradation et les instruments de protection des bandes côtières italiennes, Societá Tecneco, 1975.

In 1971, a Shore and Mountain Planning Act 4/ was adopted to promote the co-ordinated utilization of land in shore and mountain areas, with a view to preserving their natural assets and public availability, as well as to ensure that development of the areas for recreation and tourism occurs on the basis of the best interests of the community as a whole.

According to this Act, in shore areas along the coast, the erection of buildings, structures and installations may only be undertaken in conformity with approved shore plans. 5/ Shore areas include inland and shoreline areas which are of significance to the siting of holiday dwellings and tourist developments. In general, no building or construction activity is permitted on the shore within 100 m from the shoreline measured horizontally at normal high water. 6/

The purpose of the ban is to prevent the remaining open coastal areas from being covered by an unbroken chain of structures. The erection of buildings, structures or installations may only be undertaken in conformity with approved shore plans. Certain laws and regulations are also made applicable to coastal areas regarding such aspects as free access to the beaches (Open Air Recreational Act, 1957), site location for power plants and industries or coastal environment protection (Water Pollution Act 1970, Nature Conservation Act 1970, and the Neighbour Act of 1961).

The Country Governor is in charge of the preparation of a given shore plan. He issues guidelines and calls for meetings of all the known landowners involved. For the preparation of a shore plan, special powers are given to appropriate authorities to summon landowners to attend meetings to prepare draft plans which may be adopted by a simple majority. 7/ The draft plan will then be reviewed by the local building council which may return the plan for reconsideration. The draft plan also has to be posted for a reasonable period of time for public comments. 8/ Once the building council approves it, the plan is then submitted to the municipal council which may again approve or reject the plan. This process is designed to ensure better co-ordination at all levels and of all parties concerned.

Under the Urban and Rural Zones Act and the Urban Planning Act of Denmark, coastal areas are normally given top priority. 9/ Industrial areas are laid out in urban zones in conjunction with urban planning. Industrial plants (except farms, forestry and fisheries installations) are not permitted to locate in rural areas which include the coasts. There are special regulations on extraction of minerals from the beach. 10/ Landfills on the coast may be undertaken with

---

4/ Act No. 103 of 10 December 1971 concerning Planning in Shore and Mountain Areas (Shore and Mountain Planning Act) with Amendments of 7 June 1973 (unofficial translation).

5/ *ibid.*, Special programmes are made for shore plans in Sections 7-10 of that Act.

6/ *ibid.*, Section 3. Exceptions are made for special areas required for military or other approved uses (Section 6).

7/ *ibid.*, Section 8.

8/ *ibid.*, Section 10.

9/ See Landskabsanalyse for Ringkobing amt, 1972; Status 1976, published by the Conservation Planning Committee of Copenhagen, 1976.

10/ Act No. 292, 23 May 1973.

permission from the Ministry of Public Works. Prior to granting any permission, the Ministry of the Environment is to be consulted to ensure that the proposed measures are not in conflict with any aspects of nature conservation, general planning or protection against pollution. There is a Coast Inspectorate whose tasks include administration of the territorial sea, dyke legislation, coastal fortification and construction projects.

The problem affecting the coastal areas of Britain are mainly associated with the demands of industry and commerce, the needs of an increasing and more mobile population with more time, money and desire for open-air recreation (especially that which is water-based), and the growing numbers of those who wish to retire and live in close proximity to the sea. Measures have been taken to deal with specific areas and issues. A comprehensive study of the coastline of England and Wales by the Countryside Commission during 1966-70 led to recommendations that certain stretches of undeveloped coast of particular scenic beauty should be treated as "heritage coast". These measures are dealt with below in the context of nature reserves and parks. 11/

Coastal area management as a whole in Britain is regarded as one aspect of the land-use planning system. Certain planning studies of the coastline have been carried out, and development plans prepared by the local maritime planning authorities provide the basis of decision-making for all forms of development. 12/ Under the Town and Country Planning Act 1971 of England and Wales, all "development" requires the prior consent of the local planning authority. An exception is development undertaken by a government department which by law does not require planning permission. The term "development" is broadly defined as the carrying out of building, engineering, mining or other operations in, on, over or under land, or the making of any material change in the use of any buildings or land.

When deciding upon an application for planning permission, the authority must keep in mind the provision of the development plan for the area concerned and any other material considerations, as for example the effect of the proposal on road safety or the beauty of the surroundings and the effect on public services, such as drainage and water supply. 13/ If the proposals for development do not accord with the plan, the local planning authority can still give its consent if it believes that they do not involve a substantial departure from the plan or affect the whole of the neighbourhood. In the case of other departures from the plan, the authority must (if it proposes to permit the development) give public notice of the application, ask for representation, and must send a copy of the application to the appropriate minister whose interest may be affected. However, he does not normally intervene unless it appears that important planning principles or issues of more than local significance are involved.

After considering an application for planning permission, the local planning authority can refuse its consent, grant unconditional permission, or give its consent subject to such conditions as it thinks fit (it may require, for example, the new use of land to cease, or the building or works to be removed at the end of a specified period). There is a right of appeal to the minister responsible against refusals or conditions attached to a grant of permission.

11/ See Chapter 11.

12/ Information received from the United Kingdom, dated 5 March 1976.

13/ *ibid*.

There is an important series of acts and procedures, other than those concerned with planning, designed to protect the environment and to supplement the planning system. In many instances, conformity with the requirements of these acts and procedures is necessary before development can proceed. 14/

The reclamation of derelict industrial land is assisted by government financial grants to local authorities for restoring and brining into positive use the abandoned sites. Outside the development and special development areas, an application for planning permission that will lead to the creation of industrial floorspace of an area greater than 5,000 ft^2 in the Southeast, and greater than 10,000 ft^2 elsewhere, must be supported by an industrial development certificate.

In Kenya and Tanzania, coastal area planning falls within town and country planning legislation. 15/ Provisions are made in the legislation for the establishment of planning institutions for local areas so that local characteristics can be considered. Machinery exists in each country for the evaluation of planning schemes. Tanzania has an independent body i.e. the Planning Control Board, to supervise and co-ordinate all development projects of the country, including those proposed by the government itself. Since the government is itself a developer and in order to ensure that all planning requirements are taken into account, the Government Director of Town Planning is an official member of the Board.

The Tanzanian legislation indicates what matters may be provided for in the planning schemes, 16/ for example (i) the location, density and reservation of land for open space, roads, car parks and bus stations; (ii) dwelling houses, flats and tenements of various classes and densities; (iii) shops, offices, trading premises and other commercial development with or without residential accommodation; (iv) zones or areas in which development of any kind is restricted or prohibited for any period; (v) the size, height, or spacing of buildings; (vi) industries of various classes, warehousing and service trades or any particular industry or trade. These regulations are equally applicable to coastal areas.

Kenya's legislation is very general on this aspect and leaves much discretion to the planning authorities. The Development and Use of Land Planning Regulations (1961) under the Land Planning Act simply require that an application for development show the existing development in the area, as well as proposed roads, density zones and the area in which subdivision is permitted. No other requirements are specified. Regulation 11(2) requires that an application indicate the part of the land which the applicant intends to surrender for "public purposes", e.g. educational, medical and local government purposes. This regulation may thus be used to improve the conditions in the area in which development is intended to be carried out. The government and the planning authorities presumably have

---

14/ For example, the Dumping at Sea Act 1974 prohibits, except by licence, the permanent deposit of any substance or article in the sea or tidal waters from a vehicle, ship, aircraft, hovercraft or marine structure or from a land structure such as a conveyor belt. The Coast Protection Act 1949, referred to in Chapter 11, section 5, is another example.

15/ See Town and Country Planning Ordinance, Chapter 378, Laws of Tanzania; (Kenya) Land Planning Act, Town Planning Act and the Local Government Regulations.

16/ Section 26 of the Town and Country Planning Ordinance.

guidelines on this aspect of planning from which they have been regulating the development in the country. More clarity could be achieved if those guidelines were made public or were incorporated in the planning legislation.

At present, most of the buildings in the coastal area of Kenya are put up by private developers. The consequences of their activities have resulted in ad hoc development, traffic bottlenecks, overcrowded schools and shopping centres. The provision of land for basic and adequate living space and of water, sanitation, sewage and transport has been a serious problem, and better use of Regulation 11(2) is called for.

In New Zealand, coastal area planning is viewed in the context of the making and enforcement of regional and district schemes which are provided for under the Town and Country Planning Act of 1953, and is carried out by regional and district planning authorities. 17/ The Act applies to all city and suburban areas, where in general, the coastal area is significantly built-up. Most of the coastline is situated in areas controlled by a county council. Local authorities in rural areas are governed by the Counties Act 1956. 18/ Under this Act, specific powers are given regarding subdivision of land, road, minimum frontage and area requirement, water supply, sewage disposal, road formation and reserves. Of particular interest are those provisions which reserve along the sea-shore and the banks of lakes and rivers areas for public recreation when land is subdivided. 19/ The requirement of foreshore reserves (i.e. that area of land which is held in public ownership immediately above and/or below the mean high water mark (about 66 ft) for public access) has led to a substantial portion of the coastal areas being accessible to the public.

Regional and district planning involves the formulation of the objectives of the community concerned, the examination of alternative policies for the achievement of these objectives, and the selection of policies which best achieve the objectives. The selected policies are then stated in the regional or district planning scheme. It is through these means that regional or district policies, integrated with policies of central government, can be clearly stated and implemented. The 1953 Act was recently amended to declare that the preservation of the natural character of the coastal environment and the protection of the coast from unnecessary subdivision and development are matters of national importance which must be recognized and provided for in regional and district schemes.

The major technique for implementing land use objectives and policies is zoning. The type of zone depends on the objectives and policies adopted by a local authority. A number of local authorities have adopted zoning laws for their coastal areas. For example, the Code of Ordinances 20/ specifies the "predominant

---

17/ The Act and Regulations (1960) provide for comprehensive land use planning and development control at a district and regional level.

18/ This Act (and its subsequent amendments) deals comprehensively with local government functions in counties.

19/ The County Act, Section 29.

20/ It is the detailed 'rule book' for the administration and implementation of the policies set out in the scheme statement.

uses" allowed in a zone, the restrictions to which each predominant use is subjected, and the "conditional uses" which may be permitted in a zone. Conditional-use provisions are intended to cater for uses which are not suitable everywhere, but which may be suitable in some locations under certain conditions. 21/

A district scheme usually deals with:

(i) the preservation of objects and places of historical or scientific interest, or natural beauty;

(ii) the designation of reserves for national, civil, cultural, and community purposes; for afforestation and water catchment purposes; for recreation grounds, ornamental gardens, parks and children's playgrounds, or for open spaces;

(iii) the designation of land for public works or for proposed public works, differentiating between government and local works;

(iv) the designation of land or buildings used for purposes of value to the community but not intended to be owned by the Crown, the council, or any other local authority;

(v) control of subdivision, including restraint on unnecessary encroachment or urban development on land of high actual or potential value for production of food;

(vi) control of development in areas likely to be affected by earthquake, geothermal activity, flooding, erosion, landslide;

(vii) access and parking space, transport terminals, and public transport systems, sewerage, drainage, rubbish disposal, lighting and water supply;

(viii) land subdivisional standards in relation to any permitted use and minimum site areas, dimensions in relation to uses of land and buildings, and finally the period during which a stage of development is to be undertaken.

It may readily be seen that a district scheme covers to a certain degree the concerns in coastal areas and provides a basic framework for action.

Although a comprehensive land planning process exists in New Zealand, there is, however, no overall control of the use of coastal waters and the seabed. Legislation such as the Harbours Act 1950 is only concerned with controlling development, e.g. reclamation, erection of structures and removal of sand and shingle, and does not provide a system for the formulation and periodic review of demands and their effect on the resources. 22/ Often, the responsibilities under the different acts are not intergrated together; this is the situation between land and water, between the Town and Country Planning Act and the Harbours Act. For example, the jurisdiction of the 1953 Planning Act stops at the mean high-water mark, and thus reclamation can be authorized under the Harbours Act prior to their landward implications being examined. Reclaimed areas only become part of the planning district once they have been approved. Structures on the foreshore can

---

21/ See Town and Country Planning Act 195, Second Schedule and Regulation, 15 (4).

22/ The information referred to here is based on communications received from the Government of New Zealand, dated 24 November 1977.

also be approved without prior planning approval. It has been suggested that the Town and Country Planning Act and the Harbours Act should be reviewed to avoid any gaps which may exist to provide a more effective system to help resolve these issues. 23/

The relationship between the Town and Country Planning Act and the Water and Soil Act is another example. In the past, land use decisions have not taken into account their impact on the quality or use of natural water, or their impact on the coastal marine environment. In many cases, co-ordination is needed among the competent authorities to integrate their different responsibilities. Legislation has been passed to effect local government reform and to create effective regional government to remedy the fragmented nature of local government. A further aspect of this is the need to develop a policy at national level which provides a framework within which regional and local authorities can plan.

Reclamation of land in coastal areas generally falls within the Harbours Act of 1950 and is subject to special procedures in order to ensure proper control and management. 24/ The procedure for approving reclamation varies according to its purpose and size. Those in excess of 10 acres in area may only be authorized by special act of Parliament. Such an application must first be made to the local Bills Committee, which receives a submission of bills from interested persons and bodies. The bill is then referred back to Parliament where it receives a second and third reading. It may be debated in the House, and following this may either be enacted or abandoned. With the introduction of environmental impact procedures, an impact report is required to accompany any local bill and forms part of the submission. For a proposed reclamation of 10 acres or less, an application may be made for an Order in Council by the Governor-General. The procedures include consideration of public rights of navigation and the public interest by the Ministry of Transport before a recommendation is made.

23/ *ibid.*

24/ The Act distinguishes between reclamation for agricultural purposes and reclamation for harbour "works" or for any other purposes such as industrial land, marinas or recreational reserves.

CHAPTER 11

Measures for Protection of Shorelines

Legislative measures have been introduced in many countries regarding the protection of shorelines. These measures are intended to deal with problems of erosion, nature reserves, water quality, discharge of effluents, and oil pollution. These measures can be conveniently grouped under the following headings: (1) nature reserves and parks; (2) protection of shorelands; (3) water management; (4) marine pollution and (5) oil pollution.

1. Nature Reserves and Parks

In Europe, special legislation has been introduced in various countries for the protection of marine flora and fauna in general, by the creation of parks or reserves, and natural, historical and archaeological sites. There are laws and regulations regarding maritime hunting and coastal fishing, for the protection of forests and dunes, and with respect to cetain engineering works (building dykes), extraction of sand and gravel and drilling for minerals, nearshore or on the continental shelf, and for the preservation of wetlands and coastal marshlands. In Norway for example, the Ministry of the Environment is responsible at the national level for the formulation of environmental protection policy, the establishment of guidelines for pollution control, physical planning, conservation, and recreation. The Ministry is advised by various expert groups such as the State Pollution Control Authority, the State Council for Open Air Activities and the State Council for the Conservation of Nature. The overall principle is that planning should be done at the national level, whereas its implementation is aimed at the local level.

In France concern for the coastline prompted the French Interministerial Committee for Regional Development and Planning to create in 1973 a study group on the "long-term perspectives of the French shoreline". 1/ This group prepared a report and recommended certain measures be taken. One of the group's findings was that while there is sufficient law which can be brought into force for coastal managment, a major problem seems to be that of actually applying the regulations.2/ Between 1973 and 1975, a quasi-independent governmental body, the "Conservatoire du littoral et des rivages lacustres",

---

1/ The report of this group together with technical annexes are produced in Littoral français, Documentation Française, Paris 1973, 266 pp.

2/ ibid., and a speech given by President Valery Giscard d'Estaing in Vanues on 8 February 1977 on "Politique nationale d'aménagement et de protection du littoral".

was created, the purpose of which is to acquire saltwater shoreline property (or property which lies around freshwater lakes over 1000 ha (2471 acres) in size), in order to protect such land from urban encroachment, to protect the "ecological balance" of chosen areas or to enhance public access to the shoreline.3/ An initial 5 m francs was apportioned in 1976 for that purpose. The sum has increased to 10 m francs for 1977-78. As of 1977, a total of 3274 ha have been acquired under the fund.

There is also a National Administrative Council which is composed of ministerial representatives (12), representatives of shoreland Councils (10), members of Parliament and Senators (5) and "qualified individuals" (3). The Council administers not only the fund but also defines its policies. This National Administrative Council is the only major institution in France whose sole objective is intervention in the process of coastal development. At the subnational level, there are five regional shoreland Councils, each of which consists of elected representatives (between 12 and 18) from surrounding planning regions. The shoreland Councils are purely consultative in nature, but are supposed to make recommendations concerning proposed and desirable land acquisitions.

In England, under the National Parks and Access to the Countryside Act 1949, areas of ecological importance, which include beaches, dunes, swamps and marshes, can be safeguarded by being declared National or Local Nature Reserves. The Nature Conservancy Council has a duty under the Act to notify the local planning authority of any "sites of special interest" by reason of its flora, fauna or geological or physiological features. About 70 coastal reserves out of 153 national nature reserves have been declared in England, Scotland and Wales to protect rare species of flora, peat islets or limestone.4/

Comprehensive studies of the coastline of England and Wales conducted through the National Park Commission (which became the Countryside Commission) during 1967-70 resulted in the collection of a great amount of information. Recommendations were made that certain stretches of undeveloped coast of particular scenic beauty should be treated as "heritage coast".5/ Many parts of the coasts (over 400 miles) are now acquired and managed by the National or local Trusts, which are bodies independent of government. Trust

---

3/ It was set up by Law No. 75-602 of 10 July 1975, Journal Officiel 11 July 1975, discussed in D. Legrain, "Le conservatoire de l'espace littoral", Espaces, Vols. 15-17, 1975, pp. 53-58.

4/ For details, see J. A. Steers, "Saving the Coast: The British Experience", Coastal Zone Management Journal, Vol. 3, nos. 1 and 2, 1978, pp. 7-24, p. 19.

5/ The Planning of the Undeveloped Coast, Circular 12/72. Department of the Environment, February 1972. In order to establish a baseline for measurements, the high-water mark of mean tides as shown on one inch to one mile ordinance survey maps, was selected to represent the "coastline". Offshore islands other than small rock stacks and inlets (as arms of the sea) were included in the Commission's measurements. For Scotland, however, a ferry point is not used to mark the limit of the inlets; inlets and estuaries are measured where they are over one quarter of a mile long. For measurement purposes, the "coastal belt" was taken to include all levels, whether or not it is intermittently covered by water, within one mile to landward of the arbitrary datum line. Land lying between the high-water mark is excluded.

ownership means that nothing less than an act of Parliament can bring about any change, and thus ensures the status of these special coastal areas.

In Scotland, while oil is being exploited, a coastal nature reserve has been established near the Nigg and Udale Bays, covering an area of 750 ha of international importance to migratory wildfowl.

The Conservation of Nature Act of Denmark makes provision for the establishment of beach protection zones everywhere in Danmark and the establishment is enforcible by local conservation boards. The Act provides protection against construction and landscape changes in a 100 m coastal zone. Recently, the Ministry of the Environment has drawn up new guidelines for the protection of coastal landscapes, prohibiting the building of summer houses within a protected belt of 1 to 3 km from the coast. The specific distance depends on the particular characteristics of the stretch of coast.

On a similar basis, Italy, France, Sweden and Greece have also used their legislative power to set up pro natura, maritime hunting reserves, or aesthetic forests.6/ In Sweden, there are about 900 nature reserves (the largest being 593 ha). Major tasks in these reserves are to keep protective watch over the reserves while other areas are allowed to develop without hindrance, to ensure that lumbering and agricultural techniques are consistent with natural conditions, and to check and limit forms of exploitation so as to reduce damage. In France, the Minister of Culture and the Environment and the Secretary of State for the Environment have the power to issue decrees declaring maritime hunting reserves. 7/

In Cyprus, a "white zone" may be declared for protecting places, including the coastal area, of natural beauty or special value.8/ Guidelines may also be provided regarding the choice of sites for development and for the promotion of planning control in such areas. Once a 'white zone' is declared, existing uses are to remain and any further development must be limited to essential needs, e.g. agriculture, forestry, fishing or archaeology. The order setting up the zone must be accompanied by a map together with a statement of the principles governing any development activities in that zone. A Preservation Order may also be issued for any particular building, or any area of a special soil, architectural, historic or other interest or of natural beauty. No changes are allowed in the preservation area.9/ Violators are subject to fines, and unauthorized developments subject to removal.10/

6/ For example, under the Protection of Nature Conservation Law (Greece, No. 996, 1971) the President may decree areas as "aesthetic forests". In France, the Minister of Culture and the Environment can establish maritime hunting reserves.

7/ See G. Tendron, Development of Leisure Areas and Nature Conservation in Coatal Regions and by Lakes and Rivers, paper submitted to the European Technical Conference on Leisure and Nature Conservation, Hamburg, June 1975, published by Council of Europe, 1976.

8/ See Town and Country Planning Law, 1972, Section 35.

9/ ibid., Section 38.

10/ ibid., Sections 46-55.

In the Republic of Korea, a coastal conservation zone may be established wherever it is necessary to prevent damage to the coast. Activities such as stone quarrying, clay and sand pitting, wood cutting, building, remodelling, expansion of any facility, or any geological alteration of land pitting or landfilling are prohibited in such zones. The Marine Reserves Act 1971 of New Zealand provides for the setting up of marine reserves in the sea and foreshore areas as the habitat of marine life. Such marine reserves are administered and maintained to ensure that they are preserved as far as possible in their natural state. The emphasis of the Act is to retain areas of seabed for their scientific and educational value, and it sets out machinery for the declaration of a marine reserve. Application for a declaration may be made by any university, national parks authority, any body which has been appointed to administer land having a frontage on the sea coast, or any incorporated society or body engaged in scientific study of marine life or natural history.

A coastal reserves survey was recently completed in New Zealand.11/ The purpose was to identify areas which should be preserved for their outstanding natural, historic or scientific features or for their recreational potential. The results of the survey led to the taking of administrative action to safeguard land from conflicting use or to bring it into public ownership or use, and to assessing allocation of responsibility of central, regional or local government. On the basis of this survey, detailed local plans for implementation are being made.

National parks have been set up in the coastal areas of Tanzania. 12/ Efforts to start marine parks have met with considerable difficulties. National Park officials have already identified the areas for marine parks, and have held a number of discussions with the Fisheries Division on how to implement these marine projects; outside agencies have offered financial assistance for the necessary capital development. Dynamiting of the coral reefs has caused considerable damage and is not prohibited under the law at present.13/

Kenya has established three marine parks under the Wildlife Conservation and Management Act 1976 14/: Malindi Marine National Park (covering the waters of the Indian Ocean within Kenya's territorial limits); Watam Marine National Reserve (comprising the Mida Creek); and Kisite-Mpunguti Marine National Park (south of Wasini Island, covering some 21 km^2 and encompassing three offshore islands). All the marine life in the parks is protected and hunting is prohibited unless licensed.

For many years, Argentina has enjoyed a national park system.15/ Provincial authorities may also establish provincial parks.16/ The use of

---

11/ See The Coastal Heritage, publication by the Department of Lands and Surveys, New Zealand, 5000/P.82/1975.

12/ For example, the Mikumi National Park and the Setons Game Park.

13 See Annual Repot, Tanzania National Parks, 1975, p. 14.

14/ Chapter 376, Law of Kenya.

15/ Law 12,103, 29 October 1934.

16/ Decree 2,091, 2 May 1969.

this power in the coastal zone is shown by a 1969 decree of the Buenos Aires Province which declared the Martin Garcia Island in the La Plata estuary as a provincial park and recreational reserve. Hunting of wild animals in national territory or any other territory under national jurisdiction is forbidden.17/ Under certain conditions, however, licences may be issued for sports hunting, while permits for commercial hunting are limited to those species specifically designated by government as suitable for this purpose.

The Barbados Parks and Beaches Commission Act gives the Commission the power to make regulations to establish sanitary conditions, practices to be observed in respect of public parks and beaches, and measures for the preservation of beaches. Under the Act, beaches cover not only the foreshore but also the first 100 ft of the landward limit of the foreshore.

In the last ten years, legislative acts have been adopted in Yugoslavia for the protection of fauna and flora both by the Federal Government and by the Union Republics.18/ Measures taken at the Adriatic coast are examples of harmonizing development, e.g. cultivation of land, and tourism on the one hand and the setting up of green areas and parks to preserve historical settlements and monuments on the other.

In Poland, natural reserves may be created by Executive Acts of the Minister for Forestry. In the shore regions, a whole range of reserves has been created. For example, the reserves "Kadynski Las" in the Gdansk voivodship and "Bialoglowa" in the Gdansk voivodship were specifically created for the purposes of research, training and the preservation of a valuable part of an old forest. In the reserves, no felling of trees, diverting waters, erecting buildings or structures are permitted.19/ Similar reserves can also be established in Romania under the Law on Environment Protection of 1973.20/

In many South Asian countries, coasts are places in which shrines and other architectural structures of archaeological value exist and where ancient treasure may be discovered, especially during development works. There are laws regulating finds, monuments and archaeological sites.21/

17/ The legal position of Martin Garcia Island was changed after the 1969 Treaty between Argentina and Uruguay.

18/ For example, the Slovenian Law on Protection of Nature, 1971.

19/ Regulation of the Minister of Forestry, 23 January 1973.

20/ Lois sur protection de l'environment, Revue Roumaine des Sciences Sociales, 1975, No. 21, p. 59; Decree No 76. of 1953, Decree No. 43 of 1954.

21/ For example, West Bengal Preservation of Historical Monuments and Objects and Excavation of Archaeological Sites Act, 31 of 1957; Orissa Ancient Monuments Preservation Act, 12 of 1956; Andra Pradesh Ancient and Historical Monuments and Archaeological Sites and Remains Act, 7 of 1960; Madras ditto, 25 of 1966; Kerala Ancient Monuments and Archaeological Sites and Remains Act, 26 of 1969; Mysore Ancient and Historical Monuments and Archaeological Sites and Remains Act, 7 of 1962; and Gujarat Ancient Monuments and Archaeological Sites and Remains Act, 25 of 1965 (the authority has the power to purchase lands and antiquities).

2. Protection of Shorelands

Australia has a Coast Protection Board established under the Coast Protection Act 1972. The purposes of the Board are to protect and restore the coast from erosion, to develop any part of the coast for aesthetic improvement or for more appropriate use, and to conduct research on protection, restoration or development of the coast. The Board is preparing a management plan for South Australia's metropolitan beaches. The plan considers land use, land tenure, physical features, public access, foreshore and beach activities and measures to meet the problem of erosion and pollution. The plan takes the view that erosion problems form only one part of the wider management problem of a coastline, and stress that all coastal activities, such as harbours, residential, recreation, conservation, industrial or beach reservation, are closely interrelated. The solution of erosion problems might be solved by means other than engineering works and local authorities are advised to consider immediately preventive measures to avoid erosion problems, rather than to wait for engineering solutions.22/

Under its Beach Protection Act, beach erosion control districts are established on the coast up to 1600 m offshore from low-water mark and up to 400 m inland of high-water mark. The guidelines established for beach protection suggested that such districts should extend the full 1600 m offshore, and the inland boundary should be drawn to include all areas considered to be vulnerable to long-term erosion over a 50 year period. Coastal protection in Australia is further strengthened by the Water and Soil Conservation Act, one of whose purposes is to promote soil conservation and prevent damage by flood and erosion.

The protection of soil against erosion is one of the most urgent problems in Bulgaria in view of the great quantities of shoreland being washed into the Black Sea. Several statutes are applicable to the protection of shoreland and soil against erosion 23/ and come within the competence of the Ministry of Forestry, Ministry of Agriculture and local People's Councils. The relevant laws require the competent authorities, for example, to plant trees as a means to control erosion.

The Law on Protection of Air, Water and Soils against Pollution deals also with sanitary aspects in tourist and health resorts and the use of drinking water. Agricultural land is protected through a system of high tariffs from being unreasonably diverted to non-agricultural uses. The funds obtained are channelled into land improvement, drainage and other purposes aimed at increasing agricultural production.

In the Soviet Union, shoreland protection is of great importance in the Black Sea region. For example, the Law on Protection of Beaches on the Black Sea 1962 prohibits the removal of pebbles from the beaches and feeder streams. The Law on Urgent Measures for Protecting the Black Sea Coast against Destruction and for Rational Utilization of the Resort Territory of the Black Sea Coast 1969 provided that no new factories or buildings could be built within 3 km of the coast. Fundamental rules on land protection,

---

22/ For the purpose of the plan, a working definition of the coast is given and covers from the 100 m landward line measured along high-water mark to a seaward limit of 2 nautical miles from the low-water mark. At the same time, it is emphasized that ocean outfalls, offshore mining and dredging in that area should also be included.

23/ For example, the Law on Protection of Air, Water and Soils against Pollution, the Law on Protection of Agricultural Areas, the Forest Law.

including shorelands, are contained in the principles of land legislation. Soil erosion is regulated by the Council of Ministers' decision on urgent measures for the protection of soil from wind and water erosion of 20 March 1967 24/ and by legislation within the republics. A general scheme of an anti-erosion programme has been instituted in the Ukraine SSR. 25/

Under the British Coast Protection Act 1949, maritime district councils are designated as coast protection authorities and have powers to carry out works. They have a general responsibility for the protection of the coastline(except certain estuary waters) against erosion or encroachment by the sea. Schemes for coast protection are submitted to the Secretary of State for approval and loans and grants may be given. Water authorities are responsible for the defence of low-lying land against flooding, whether by inland water or by the sea. Applications for licences for offshore dredging of sand and gravel from the seabed are submitted to the Crown Estate Commissioners, who consult with other government departments before any licence is issued.

India is aware of the need for soil conservation, particularly in areas open to erosion such as Bengal and Kerala. The law is concerned mainly with two problems, the question of land ownership for taxation purposes, and protection against flooding. 26/ Some statutes specifically authorize embankments to be built, maintained and protected. The control of such matters has always lain with government.27/ The basic principles common to this type of legislation are (i) the identification of a body responsible for detecting where works are needed, and for carrying them out; (ii) the empowering of this body; (iii) conferment of supplementary or ancillary powers; (iv) penalties for interference or infringement. Thus the legislation sets up a legal framework, while the rules promulgated within that framework set up an administrative framework. Of special interest is the Kerala Land Development Act, 28/ which may well be adapted in regions with similar erosion problems.

The purpose of the Act is to prepare and execute schemes for the conservation and development of oil resources, the control and prevention of soil erosion and the reclamation of waste lands. There is a Land Development Board which prepares plans. The initiative lies with local politicians, whose constituents or interested parties may press them for improvements or the initiation of new ideas. The Board conceives plans and devises schemes to free the local politicians from responsibility for detailed work, such as feasibility studies. Then there are the District Committees composed of personnel recruited locally on a part-time basis. They are the actual initiating functionaries. The District Committees make recommendations to the Board. The Board's schemes may provide for such matters as reclamation, prevention of soil erosion, and drainage of canals, streams and rivers. The financing of the authorized work is to be borne by the owners of the land affected. In this manner, the schemes automatically

---

24/ SP SSSR, 1967 No. 9 p. 45.

25/ Vedemosti verkhovnogo Soveta SSR, 1960, No. 23, p. 175; SP USSR, 1967, No. 5, p. 47

26/ See for example, the Bengal Alluvion and Diluvion Act; Bengal Alluvial Lands Act, 5 of 1920.

27/ For example, the Sunderbans Act 4 of 1915.

28/ Kerala Land Development Act, 17 of 1964, amended by 16 of 1973.

involve the owners in the work of the District Committees. The taxpayers are partly relieved of the financial burden. To strengthen the system, government is given the power to reserve the right to carry out work as devised by the Board and to recover the cost from the owner. The Board and its employees have rights to enter, inspect and survey the area involved.

Argentina has established a Council to develop a plan for soil improvement and cultivation in the federal territories of Tierra del Fuego, Antarctica and the Islands of the South Atlantic. 29/ Concerns about soil erosion led Brazil in 1975 to order the Ministry of Agriculture to designate those lands where special protection of the soil is needed. 30/ The owner of such land must begin the corrective work to prevent soil erosion and must complete such projects within a specified period. No credit for such land may be given by government lending institutions for other purposes until the projects have been completed.

A coastal conservation law in Sri Lanka has been under consideration by the National State Assembly. 31/ The introduction of this legislation was in response to the need to combat coastal erosion, particularly along the southwest coast where several miles were lost because of sea erosion. While coastal erosion in Sri Lanka forms part of the natural process because of the tilting of the island, the situation is seriously aggravated by sand and coral mining and by the construction of outlets for drainage. The pollution of coastal waters by human and other wastes degrades water quality and seriously affects the growth of natural reefs. A new division under the Ministry of Shipping, Aviation and Tourism is being organized for the planning and development of the coastal area and for the effective implementation of the proposed legislation. The Director of Coastal Conservation is empowered to declare certain areas of sea as 'prohibited areas' for the dumping of any soil or material. Certain coastal areas may be excluded from excavation of materials from or under the sea-shore.

Brazil enacted in 1975 a law requiring industries to take measures to prevent pollution and to correct damage caused by contamination of the environment. 32/ 'Industrial pollution' is defined as anything which will cause harmful changes in the environment and which may injure the health, safety, or well-being of the people, the flora and fauna, or any other natural resource. The Ministry of Interior is to issue the necessary regulations and within the areas of their authority, the state and municipal governments may also make rules and take measures to prevent or correct industrial pollution. However, industrial establishments 'of high interest' 33/ to national security or economic development may be given special exemptions. Both the federal and state government may adopt emergency measures to reduce pollution; Rio de Janeiro and Sao Paulo have

---

29/ Decree 8 of 5 January 1970.

30/ Law No. 6225 of 14 July 1975.

31/ See the Gazette of the Republic of Śri Lanka, Part II of 30 July 1976, Supplement, Coastal Conservation. The purposes of the law are very broad: to consolidate the law relating to coast conservation and the administration, control, custody and management of the coastal zone and to make provisions for matters connected therewith or incidental thereto.

32/ Diario Ley 1413, 14 August 1975.

33/ The term is not defined. Presumably these would include petroleum and electricity.

been declared 'critical areas'. In approving the location of industrial projects, government agencies are to avoid aggravating the situation in critical areas by granting incentives for private investment.

3. Water Management

In Yugoslavia, the sea and the seacoast are treated as a part of the 'commonwealth' and as such are specially protected. They can be used only under conditions prescribed by law in order to ensure their rational exploitation. Jurisdiction over the waters and the coast is essentially in the hands of the republics and autonomous provinces. The federation retains powers concerning the overall management of waters involving two or more republics, international waters, and Yugoslav coastal waters as regards pollution control. The basic management of water courses is governed by a master plan drawn up by mutual agreement among the Republics concerned. The classification of international and interrepublic waters is established by a federal ordinance which prescribes, inter alia, maximum permissible concentrations of noxious substances in the water. The republic and provincial water-laws lay down the basic principles concerning utilization, uses of watercourses, flood control, water conservation and erosion control. These laws also deal with questions of implementation, supervision and sanctions.

Water management communities (or ventures) are set up to deal with control and supply of water in the country, including the coastal region. These communities are formed by citizens and working people, and have a wide range of authority; from planning, execution of plans, maintenance and control of water regimes, watercourse regulation, training, and development of navigation waters. As for irrigation, drainage, hydroelectric power and industrial water supply, there are specialized communities (or ventures) formed by users and suppliers. The law of the Republics authorizes local government to prescribe conditions under which ventures can be organized. For example, irrigation and drainage ventures are formed by participants from the enterprises maintaining the system, and citizens and enterprises which use the land improved or reclaimed by the system.

Water authorities exist at federal, republic and provincial levels. Their functions are to draw up legislation on all matters concerning water and to enforce the relevant laws. Every water area is required by law to have a water management master plan to cover a period of 15 to 20 years. The draft is made available for public scrutiny and is submitted to the republic (or provincial) assembly for approval. Once it has been approved, it is then incorporated into laws for enforcement.

In Kolobrzeg, a coastal region in Poland, special laws are issued to protect waters needed for medical purposes. 34/ Any construction or rebuilding within the defined region is closely controlled and, in certain areas near protected waters, any exploitation of clay, gravel or waters for industrial purposes is prohibited.

Systematic sanitary control of coasts and summer resorts or ports on the Baltic sea is carried out by the Inspection Service of the Ministry of Public Health and Welfare. This control covers the total coastal length on a seashore which is usually a few hundred metres. The corresponding measurements and analysis are carried out by the laboratories of the local

---

34/ Order of the Chairman of the Higher Mining Office on Establishment of the Area of Mining Protection of Medicinal Water Springs, Mineral Waters of the Health Resort-Kolobrzeg of February 14, 1963, M.P., 1963, No. 15, Item 91.

authorities which are administratively attached to the Central Water Resources Administration. Since 1971 a comprehensive research programme has been carried out for pollution control in the central Baltic and the straits. Certain ecological parameters have been measured. The two main rivers, Odra and Vistoula, which lead into the Baltic, maintain an adequate water quality level in order to satisfy the quality standards established in the country for industrial and irrigation use. Systematic control of water pollution is carried out for rivers with basins larger than 300 km^2. The frequency of surveys depends on the economic importance and the rate of pollution increase of the rivers under study. However, the intervals should not exceed five-year periods. The Odra and Vistoula are examined every year.

In Bulgaria the use of water and water basins is set out in a well-defined water economy plan which, in turn, is governed by the Water Economy Act. The construction of sewage treatment plants to deal with urban and industrial waste waters is required under the national economic plan. Under its Water Law of 1969, it is prohibited to undertake activities or to carry out construction work which could harmfully affect the natural condition of water, the run-off, the river flow capacity, the quality of water etc., and special penalties are applied against violators. For the protection of the Black Sea against pollutants resulting from inland sources, several technical measures were undertaken in 1970, such as the treatment of sewage and effluents from housing, camping and industrial sites. The Ministry of Forestry and Protection of the Natural Environment is responsible for the enforcement of these measures. Expenditures are shared among the central authorities and establishments whose activities include the discharge of waste waters.

The Water Law of Romania 1974 requires the preparation of water economy programmes to be based on schemes for the management of water basins. The programmes are intended to safeguard rational and complex utilization of water resources in the light of social and economic development and the quality of environmental protection. General plans for each water basin are formulated on the basis of development plans. Each scheme contains an analysis of surface and groundwater potential from the standpoint of quantity and quality, and an assessment in terms of utilization of waste bilans. Emphasis is given to environmental aspects, future needs, hydroelectrical potential, soil, erosion and management methods.

Guidelines for future economic and technical policy in the utilization and protection of water resources in the Soviet Union are set out in the General Scheme for the Multipurpose Use and Conservation of Water Resources 1980, which covers the whole territory. This General Scheme provides for the distribution and development of the productive forces in the country, while taking into account the need for preserving and improving the state of the hydrosphere as one of the elements in the natural environment.

Traditionally, waters in many Latin American countries were subject to appropriation and control by private persons. Since the 1960's, there has been a shift away from the sanctity of private property rights in order to give powers to the State to manage waters and related problems. Hydraulic reserves, waterfalls, currents and natural water deposits are now in the public domain.35/ The law establishes protected water zones together with the surrounding land necessary to conserve these water resources. Hydraulic

---

35/ For example, Venezuela, Decreto Ley 1333, 6 March 1969. Some trend is also followed in Peru (General Law of Waters, Diario Oficial 17,752 of 1969) and Ecuador (1972 Supreme Decree declared her waters a national asset for public use and consequently outside commerce).

works and dams are subject to special regulations, that is they must be constructed in a manner so as to avoid affecting the aquatic animal life; dams and dykes have to be built with staircases to permit the passage of fish.

The flow of river water is relevant to the availability of fresh water for drinking, irrigation and for industrial purposes, and it affects drainage and sewerage, which in turn affect health. The powers and duties of various regulatory bodies are intimately connected with drainage and flood control. In India River Boards are established to administer and regulate interstate rivers and river valleys. 36/ The Boards are charged with advising government on such matters as conservation, control and utilization of water, promotion and operation of schemes for irrigation, water supply, drainage, hydroelectric power, flood control, navigation, afforestation and control of soil erosion, and the prevention of water pollution. Each Board also has power to acquire, hold and dispose of property, to undertake survey and inspection, to collect topographical, meteorological, hydrological and subsoil water data and to publish statistics. In Kerala flood control, drainage and paddly-rice growing are the concern of a special officer. In some regions, government has the power to deal with the non-exploitation of lands under the command of the irrigation scheme, and cultivators may be compelled to make use of the scheme.37/ In Sri Lanka government has the power to reclaim low-lying land in Colombo District for development purposes. 38/

The adoption of the Water and Soil Conservation Act in New Zealand in 1967 has led to the establishment of a policy for the conservation, allocation, use and quality of water and procedures for its control and management. The policy ensures adequate supply of quality waters for national uses, e.g. primary and secondary industries, drinking water, fisheries, wildlife habitats and recreational uses.

Provision is also made for the classification of waters suitable for various public uses. The purpose of this classification is to guide the location of waste discharges so as to minimize conflicts, and to regulate their quality, thus providing a plan for their control. The provisions of this Act also extend the powers of the Soil Conservation and Rivers Control Act to include its operation in maritime areas regarding conservation of soil resources and control of erosion and flooding.

In recent years, conversion of natural wetlands, e.g. swamps and flood plains, to productive farmland in highly developed regions of New Zealand has left only remnants of formerly large areas of natural wetlands. 39/ Recognizing the major functions of wetlands as those of flood and erosion control and nursery area for food and recreational uses, those institutions specifically responsible for the conservation of wetlands are concerned and

---

36/ See for example, the River Boards Act, 49 of 1956.

37/ For example, the (Indian) Maharashtra Act, 47 of 1965; East Bengal Embankment and Drainage Act, 1 of 1953.

38/ Colombo District (Low-lying areas) Reclamation and Development Board Act.

39/ Water and Soil Management Guidelines, published by the Ministry of Works and Department for National Water and Soil Conservation Organization, Wellington, 1975, 9 pp.

have proposed guidelines aimed at clarifying approaches and actions in managing water and soil. 40/

In the Republic of Korea, the Shoreland and Adjacent Water Space Control Law was enacted in 1961 to preserve, use and manage shoreland and adjacent water space. It also prohibits the disposal of any noxious materials and any activities which may damage the quality of the coastal environment. Any building, dredging or deforestation in coastal areas requires prior permission.

According to the Health Decree 1965 of Greece, sea waters are classified into categories according to their use, and for each use, water quality standards are established. There are also general regulations concerning the quality of waste waters discharged into the sea and the conditions under which discharge is accomplished. According to the same decree, authorization is required to discharge waste water into the sea, and it is granted on the basis of existing standard requirements and supplementary ones depending on the particular case. The competent authority for the implementation of the decree of 1965 is mainly the Ministry of Social Affairs, by means of its central and regional services. As far as technical and economic measures are concerned, the construction of municipal waste networks, waste treatment plants and new discharging installations is being promoted and will be financed through loans, government financing or other sources.

4. Marine Pollution

It is generally recognized that pollution in coastal areas has adverse effects directly or indirectly on most economic activities. 41/ Recently, the problem of coastal area pollution has become particularly serious, primarily because of the importance and function of coastal areas in the whole national economy and the changing nature of sea discharges itself, as the intense exploitation of the coastal zones continues. The importance attributed by countries to pollution of coastal waters is reflected in their efforts to prepare suitable legislation ensuring efficient pollution control. Some countries have adopted direct legal measures concerning, for example, the conditions for safe discharge of effluents into the sea and the authorization required. These measures are promulgated either in the form of general regulations, directives to specific users or in court decisions. As mentioned in the previous section, the coastal waters of some countries are classified according to their uses and for each use, water quality standards are established.

While the accelerated development of coastal areas has led to a change in the nature of sea discharges, the pollution of coastal waters does not always originate solely in those areas. A considerable proportion of the pollutants are also carried by rivers. The rivers also carry nutrients and pesticides contained in run-off and drainage water from land under cultivation. Because of the complexity of the marine environment, the problem of coastal water pollution is thus linked to that of river pollution.

The present concern in many countries has been focussed on heavily-polluted rivers and estuaries with the aim of obtaining an acceptable

---

40/ ibid.

41/ The Pollution of Coastal and Estuarial Waters, published by the Economic Commission for Europe, Committee on Water Problems, ECE/WATER/6 (mimeographed), 8 August 1974, p.3. This document provides a survey of the problems of marine pollution in 22 member countries of the Commission.

level of pollution. Little attention has been given to the management of water resources which remain relatively unpolluted. For most developing countries whose objectives are to create conditions for fast economic growth, there is a real danger that they may soon face the same kind of problems which many industrialized nations are now confronted with.

As pointed out in a UNEP report, 42/ only recently has there been any legislation aimed at the control of marine pollution *per se*. The prohibition or regulation of discharges of harmful wastes into coastal waters has been promulgated mostly to prevent the contamination of fishery resources and to protect human health. The oceans have been considered as an infinite receptive tank of undesirable matter produced by man, and consequently national legislation concerning the control of wastes into the marine environment has lagged far behind the actual requirements of control.43/ Often, such control measures are merely an extension of those established for freshwater pollution and there are areas where literally no regulation exists, such as land run-off or the atmospheric transport of pollutants.44/

How large an area is needed for pollution control has been a controversial issue. For the purpose of dealing with marine pollution, various distances have been proposed.45/ Some of these proposals are mentioned here. Our purpose is not to give any preference but to underline the point that situations vary in different countries and that the area needed to be taken into account varies according to local conditions. With respect to onshore width, the Federal Republic of Germany proposed the width of the port activities zone; Greece suggested that the width should vary according to the coast: flat land (500 m), shores and rocks (100 m); the United States proposed that the width should include the total drainage area of minor streams draining into the estuary and also that part of the drainage area of the major stream where any significant pollution from the stream is transferred into the estuary; Finland proposed 10 km, and United Kingdom 1.6 km. Norway considered that the width should depend on the degree of development, the density of population and the geological, geomorphological, hydrological and other physical characteristics of the area.

With respect to seaward limit, a wide range has been suggested, e.g. Belgium (3 km), Spain (3.7 km), Sweden (5 km), Portugal (55 km), Bulgaria (500 km). Several other countries have suggested different criteria: Yugoslavia considered that for the northeastern coast of the Adriatic, it was necessary to include the whole bay; Greece maintained that the limit depended on whether it was for open sea or for a gulf (which would need 5-10 km); the United States suggested the slope of the continental shelf and its fish yield.

42/ *Study of Offshore Mining and Drilling Carried Out within the Limits of National Jurisdiction,* Working Group of Experts on Environmental Law, Second Session, Geneva 3-14 April 1978, UNEP/WG.14/2, 23 February 1978, p.3.

43/ *Activities of the United Nations Environment Programme for the Protection and Development of the Mediterranean*, UNEP, 1976.

44/ See *Existing and Proposed International Conventions for the Control of Marine Pollution and Their Relevance to the Mediterranean,* published by FAO, Legal Office, Background paper No. 8, Rome, 1975, 29 pp.

45/ See *The Pollution of Coastal and Estuarial Waters,* op. cit., pp. 10-11.

The various administrative structures for coastal pollution control can be classified in two groups. In the first group, the responsibility for statutory control of coastal pollution is shared among different central departments and regional and local authorities, depending on their competence or on the nature of pollution (Cyprus, Spain, Ukrainian SSR), or on the place of pollution (Portugal).46/ The protection of coastal as well as territorial waters from pollutants coming from the open sea (oil spills(is generally the responsibility of the marine and port authorities. In the second group, the protection of fresh and salt water is the responsibility of a single body which, in most cases, is also responsible for the management of water resources. This body with its scientific units, research institutes, regional and local executive authorities usually takes the form of a national water board (Finland, Romania) with a co-ordinative capacity, or of a water ministry (Netherlands, Ukraine SSR) or of a central water resources administration (Poland).

The recent need for more centralized responsibility on matters of natural resources and environmental protection has resulted in the establishment in Sweden of a 'National Environment Protection Board' whose most important long-term project is research on the pollution of the Baltic Sea. In the United States, the Envirnomental Protection Agency has been established, and in Bulgaria, France and the United Kingdom, Departments of the Environment have been founded to which have been attached the regional authorities responsible for water resource conservation in river basins and estuaries.

The effectiveness of legislative measures largely depends on the administrative machinery for their implementation. As will be seen, where basic legislation is still scanty, responsibility for pollution control is shared among a number of departments, a system which does not seem to provide unity of action. In other cases, however, the adoption of overall legislation has led to administrative changes under which all the relevant powers are vested in a single body or under which such a body is made responsible for co-ordination of the activities of all departments concerned with any aspect of pollution. It is at this central level that pollution control principles are being developed, the application of these principles being decentralized in most cases at the level of the drainage basin. It should be noted that where such legislative and administrative measures have been taken, financial problems are also on the way to solution. The basic principle is that of sharing expenditure among all those interested in conserving the quality of water in proportion to their responsibility for pollution. Under this system, the best solutions for those concerned are also those most conducive to rational management of water quantity and quality.47/

In India, the Water Prevention and Control of Pollution Act 1974 established a Central Board, whose objectives are to promote the cleanliness of streams and wells, to co-ordinate activities of State Boards such as planning and training, and to set up a nationwide programme for prevention, control or abatement of water pollution. State Boards plan comprehensive programmes, inspect sewage, trade effluents, works and plant, lay down effluent standards, develop methods of treatment of sewage and trade effluents, and evolve methods of disposal of sewage and trade effluents on land. State governments may also declare water pollution, prevention and control areas. The special features of the Act are that it combines

---

46/ See The Pollution of Coastal and Estuarial Waters, op cit, p. 14.

47/ ibid., pp. 33-34.

research, expert participation with actual enforcement of anti-pollution rules, and that the same body determines policy at a high level and at the same time is responsible for execution at ground level.

New Zealand adopted in July 1974 a Marine Pollution Act which provides for the control, prevention and discharge of oil or pollutants into New Zealand waters. Powers and means are also provided for implementing the objectives. This Act also gives statutory effect to New Zealand's signing of the Convention on the Prevention of Marine Pollution by Dumping of Wastes or other Matter 1972. It is an offence to load any material for the purpose of dumping it at sea without obtaining a permit. Procedures have been established whereby permit applications are checked with the Ministry of Agriculture and Fisheries, the Department of Scientific and Industrial Research and the Department of Health.

Several Latin American countries have enacted detailed, sometimes comprehensive, legislation regarding marine pollution. For example, Mexico's Law on the Exclusive Economic Zone 1976 includes jurisdiction over pollution control and abatement within a 200 mile band. Its Environmental Pollution Law of 1971 forbids the discharge of 'waste waters' of certain kinds into the sewers, rivers or territorial seas, if such discharge will contaminate the quality, interfere with the purification of water or adversely affect the operation of the water or sewerage systems.48/ Likewise, waste waters which contain radioactive material or other harmful substances may not be discharged into the nation's waters nor permitted to seep into its soils. Works or installations may not be constructed if contaminated waste waters are not effectively treated. The Ministry of Health and Public Welfare and the Department of Water Resources are authorized to establish rules and conditions under which waste waters may be discharged into waterways or permitted to seep into the earth; these two agencies are also empowered to require the installation of purification systems. Detailed regulations to implement the provisions and to define 'waste waters' were issued in 1973.49/

48/ Law of 11 March 1971; Diario Oficial, 23 March 1971. Ecuador recently enacted a Law Relating to Prevention and Control of Atmosphere and Water Contamination (D.S. 374, 1975).

49/ Diario Oficial, 29 March 1973. Mexico's Environmental Pollution Law, however, has a much broader purpose than simply water problems. Directed against any substance which may contaminate or degrade the ecological system, this legislation covers all chemical or biological substances, such as "smoke, power, gases, ashes, bacteria", and "anything else which may be added to or incorporated into the air, water or earth and which can alter its natural characteristics or environment" (Diario Oficial, 11 March 1971). The Law also includes "thermal pollution, radioactive wastes, and noise levels which can alter the normal state of the air, water or land". Contamination is defined as anything which can harm or annoy the life, health or well being of humans or of flora or fauna. The executive is ordered to control the use of pesticides, fertilizers, defoliants, and radioactive materials, as well as the disposal of garbage, wastes and industrial residues. Substances not subject to organic decomposition, such as plastics and aluminum, are singled out for particulr attention. It is illegal to emit into the air substances which can harm human beings, flora or fauna. A decree, issued under this Law, describes in detail the volume and composition of gases which may not be discharged into the air (Diario Oficial, 8 September and 17 September 1971). The Ministry of Health and Public Welfare, with the aid of the Attorney-General, is to enforce these regultions.

To carry out the broad mandate contained in the Environmental Pollution Law, the executive is required to undertake a study and to design a plan for pollution abatement. The study should identify, classify and evaluate the types and sources of pollution, as well as recommend the rule and technology for adoption. The plan is to be implemented by the relevant ministries (dealing with water resources, agriculture and ranching, commerce and industry and health and public welfare) which are to ensure installation of the appropriate technical equipment, treatment processes and disposal methods. Environmental factors must also be taken into account in executing public works for urban development and national parks. Moreover, the government is to encourage decentralization in its planning. Violations of the Environmental Pollution Law, and the regulations issued thereunder are punishable by fines; in addition, offending factories or establishments may be shut down, temporarily or permanently.

Peru's General Law of Waters empowers the State to declare protective zones, in which any activity that affects water resources may be limited or prohibited. In case of scarcity, excess or contamination of water, the executive may also declare an emergency water zone, in which necessary protective measures must be taken. This Law applies to the waters of the sea up to 200 miles and the waters of gulfs, bays, inlets, estuaries, rivers and lagoons. Any solid, liquid or gaseous waste which can endanger human health or the normal development of the flora and fauna is prohibited from being dumped in the waters unless (i) they have been adequately treated; (ii) the condition of the receiving body of water is such as to permit natural purification; (iii) the seepage of such waste underground will not prejudice the use of such soils for other purposes; and (iv) other applicable regulations have been obeyed.

Peruvian law also forbids the discharge into the public sewer networks of corrosive wastes or other materials which may damage the construction of the sewers or render impossible the re-use of the recipient waters. No wastes whatever may be dumped into the territorial sea without prior approval of the government, and wastes going into the ocean must have been previously treated and must be transported sufficiently far out to sea. Herbicides and pesticides may not be used on vegetation in rivers without government permission. The Peruvian Law divides waters into five different categories, based on a variety of criteria. Permissible uses for each category are specified.50/

The waters of the Peruvian ocean areas are similarly divided into five classifications, based on various criteria such as colour, or the presence of arsenic or chrome. Class I, seawater, is designated as the 'beach zone' for recreational purposes; Class II waters are designed for conservation of marine life; Class III waters may be used for industrial purposes; Class IV waters are for navigation and ports; and Class V waters may be used for any other purpose. Generally speaking, nothing may be discharged into Class I water which would alter its natural characteristics. Nothing may be

---

50/ Class I for potable water after treatment with disinfectants, and for recreational and agricultural purposes without additional treatment; Class II for potable water after sedimentation and filtration and for fish and birds without further treatment; Class III for potable water and for agriculture after coagulation, sedimentation, filtration and disinfection; Class IV for irrigation of high stemmed plants and for industrial uses, provided such waters will not flow into the potable water system; and Class V for industrial use only (and with no connexion with the potable water system).

discharged into Class II or Class III waters which would alter their natural characteristics or alter their temperatures by more than 2.5°C. After a study of each individual case, the government may permit discharges into Class IV and Class V waters which may alter their temperatures by more than 2.5°C. In order to carry out the provisions of this Law, the Ministries of Agriculture and Fishing and of Health are to undertake the necessary studies and issue the appropriate regulations. These agencies are also charged with imposing sanctions for violations of the Law or regulations issued there-under.

In Brazil, a Division of Environmental Affairs was created by a Decree in 1973 within the Ministry of Interior to make technical studies and issue necessary regulations for the protection of the environment. 51/ The Division is under the supervision of a nine-man council, whose members are selected by the President. A regultion issued by this Division divided the recreational beach waters into seven categories on the basis of such elements as sewage particles and the presence of oils, grease, or other dangerous chemicals. 52/ The waters characterized as 'improper' or 'suspect' may be declared unfit for recreational purposes. The municipal, state or federal authorities may issue bulletins and post warnings about the classification of the waters; when considered necessary and proper, recreational activities may be prohibited in these areas.

An Argentine Decree of 1969 prohibits the discharge into its waters of any substances or waste which can harm the aquatic flora or fauna; likewise, it is illegal to interfere with the natural migratory movements of fish. 53/ A provincial law of Buenos Aires provides that private boat docks and floats may not be placed in the water without permission of the Water Department. 54/ In Tierra del Fuego and the Islands of the South Atlantic, the construction of dams on private property must not impede the passage of fish and must permit the preservation of fauna. 55/ Enforcement of this decree is placed in the hands of the Minister of Agriculture and Ranching who may, however, delegate such authority to the Governor of a particular region.

In Venezuela, once a sea, river, or lagoon has been declared a reserve, it is illegal to discharge into it any petroleum, oils, ashes or waste dangerous to aquatic fauna or prejudicial to fishing activities. 56/ There is also a general organic law whose purpose is to conserve, protect and defend the environment, and which is applicable to the question of marine pollution in general. A National Environmental Council is established to carry out the objectives of this law.

Syria has Standing Instructions for pollution control measures specifically in ports; they were issued in 1973 by the Directorate of Port Affairs of the General Maritime Department. These Instructions deal with the ascertainment of pollution, including that from port installations and undertakings, and the determination of those responsible; the immediate

---

51/ Decree 73,030, 30 October 1973.

52/ Portaria, 9 December 1974.

53/ Decree 1234, 19 March 1969.

54/ Law of Buenos Aires 7169, 10 December 1965.

55/ Decree 1284, 19 March 1969.

56/ Fishing Law, 6 October 1944.

measures to be taken in the event of pollution cases being confirmed; the subsequent measures to be taken by the port authorities; the responsibility of a port manager (having been informed of a pollution case originating from an installation or undertaking) to conduct the necessary investigations and to ensure that the case is reported; the tasks and duties of the Primary Commission for Estimating the Damage Caused by Pollution (referred to in Section 9 of the 1972 Law); and the application of procedures for dealing with cases of pollution (the necessary works are undertaken by the Pollution Control Section of the General Maritime Department, in accordance with the directives of the Director-General of Maritime Affairs and under the supervision of the Commission referred to above and the port authorities). The final clauses of the Instructions are devoted to the sums to be paid by polluters and in determining the charges to be imposed, account is taken of (i) damage to public property, including such amenities as tourism; (ii) damage to private property, e.g. cultivated land, fishing nets, and maritime equipment; and (iii) the real costs of eliminating the pollution, including compensation for the work of the various commissions concerned. Government departments, committees, ports and other agencies having jurisdiction over the control and prevention of pollution are responsible for the enforcement of compliance of the Instructions.

In Spain, a number of items of legislation deal specifically with the prevention of coastal polllution. On the administrative side, a Commission for the Prevention of Marine Pollution was set up by a Ministerial Order of 13 June 1962; its terms of reference were to include problems associated with pollution caused by the direct discharge of industrial effluents into the sea. A specialized committee of the Interministerial Commission for the Environment has been set up to deal with the control and prevention of marine pollution.

On 27 May 1967, an Order was made prohibiting the discharge by industrial installations of all kinds of petroleum products, or wastes containing such products, into the sea. This prohibition applies whether the products concerned are persistent or not. Applications for licences to discharge effluents into the sea must be submitted to the Ministry of Public Works, and must indicate the measures that are to be taken to avoid water pollution. The applications must first be approved by the Merchant Marine. An Order of 21 August of the same year specifies the materials and equipment to be kept by petroleum refineries and petrochemical plants having oil terminals on the coast, for the purpose of dealing with any accidental oil spills.

In view of the complexity and fragmentation of the legislation pertaining to coasts and of the fact that some of the provisions were mutually contradictory a single law on coasts (No. 28/69 of 26 April 1969) was introduced.57/ This Law deals only incidentally with the questions of coastal pollution, item 4 of Section 10 empowering the Ministry of Public Works to grant licences for the discharge of wastes, after prior consultation with the Ministries of Shipping, Trade, and Information and Tourism, as well as the municipalities concerned. If hydrocarbons or industrial wastes are to be discharged, the Ministry of Public Works must obtain the approval of the Ministry of Trade.

Detailed provisions are contained in the Provisional Rules for the design and construction of installations for the treatment and discharge of waste water into the sea along the Spanish coast, approved by a Resolution of 23

57/ Enciclopedía Jurídica Española, 1970 Appendix, p. 870

April 1969 of the Dirección General de Puertos y Señalas Marítimas. The first such Rules, which had been drawn up by the latter agency in 1962, were based essentially on the principle of dilution, a principle whose validity was confirmed by the experience acquired over a period of four years. At the same time, however, it was found necessary to incorporate certain bacteriological criteria. The Rules were therefore updated in 1966. Although still only provisional, the preamble to the current Rules emphasized that they are now obligatory in character.

In Egypt, the basic items of legislation that deal with water pollution control comprise the Law No. 93 of 1962 58/ on the discharge of liquid wastes and the Regulations for its implementation issued in 1967. A Higher Commission on Waters was established by a 1966 Decree; its functions include assisting projects for the discharge of human and industrial wastes and other pollutants, with a view to preventing the pollution of surface waters and groundwater.

France 59/ has adopted a multi-faceted approach to the problem, the central item of legislation being Law No. 64-1245 of 16 December 1964 concerning the administration and classification of waters and the control of water pollution. The overall policy represents a combination of procedures aimed at (i) prohibiting or, as appropriate, limiting the discharge of pollutants, and (ii) improving the quality of receiving waters. Furthermore, an inventory (subject to regular revision) is made of all surface waters on the basis of specified physical, chemical, biological, and bacteriological criteria. Numerous ministries are concerned with various aspects of pollution control, but their activities are now co-ordinated by a specialized ministry, viz. the Ministry for the Quality of Life. Other major bodies at the national level include the Interministerial Commission on Water and the National Committee on Water. Economic measures to promote pollution control have been taken, notably through the intermediary of the Intervention and Action Fund for Nature and the Environment. The 'polluter-pays principle' has been adopted, the amount of levy payable by polluters being proportional to the quantity of pollutants discharged into the water. As a result of a decision of the Conseil d'Etat, it has become possible for levies to be imposed on undertakings effecting discharges into the sea. However, the system is not solely punitive, since there are financial inducements offered by the State to polluters who undertake appropriate water treatment project.

In Turkey, 60/ the principal current measures for dealing with marine pollution are incorporated in a 1971 Law on Water-derived Resources and in the Regulations for its implementation. However, a draft law on the prevention of pollution and the protection and inspection of waters has been under consideration. It would appear that this Bill is the first in the Mediterranean area to incorporate the 'zero-emission' principle, although no timetable has been fixed for attaining this objective.

58/ Protection of the Mediterranean Sea against Pollution from Land-Based Sources: A Survey of National Legislation", UNEP and WHO, Geneva (1976), p. 7. See also The State of Marine Pollution in the Mediterranean and Legislative Controls, published by FAO, Rome, September 1972. It contains summaries of relevant legislation of the countries bordering on the Mediterranean.

59/ ibid.

60/ ibid.

In Cyprus, under the Foreshore Protection Law, as subsequently amended in 1961 and 1964, the District Officer is empowered to prohibit or impose restrictions and conditions upon the dumping of any rubble, rubbish, sweepings, litter, night-soil, engine-oils, or any other fluid lubrication oil, dry or liquid ballast, or other refuse, on any specified part of the foreshore, into the sea within a specified distance from low-water mark, or from any pier, wharf, quay, or jetty.

In the United Kingdom the discharge of deleterious substances into rivers from industry and sewage wastes has to be authorized by the regional water authorities, who control all discharges into waters. The Dumping at Sea Act 1974 prohibits, except by licence, the permanent disposal of any substance or article into the sea or tidal waters from a vehicle, ship, aircraft, hovercraft or marine structure or from other strucures such as a conveyor belt.

The Convention on the Protection of the Rhine against Chloride Pollution has been signed by most of the riparian states. This instrument gives details of how targets for reducing discharges into the Rhine might be achieved. A plan to inject salt and thus cut down pollution from the Alsace potassium mines is to be opened near Mulhouse. Financial contributions to the plan are made by France, Federal Republic of Germany, the Netherlands and Switzerland. Denmark issued a special law to protect the Baltic Sea from marine pollution. 61/

In Portugal, the Law of 1971 enforces the existing measures for the control of pollution of coastal waters. Decrees have been issued to establish an interministerial committee to study the uses of the seabed, with fines to be imposed on violators, etc. Furthermore, various regulations and measures established by private organizations aim at formulating an efficient policy for sea water protection. New legislation is on the way for the classification of coastal waters into various categories, the definition of quality standards and the setting up of conditions for waste disposal. The administrative organizations for the protection of coastal areas include: The Interministerial Committee for the Peaceful Uses of the Sea-Bed (Ministry of Foreign Afrairs); the National Anti-pollution Committee (Ministry of Navigation); the General Directorate of Hydraulics (Ministry of Public Works); the General Directorate of Health (Ministry of Health and Public Welfare). Control is practised and fines are imposed by the regional and local organs of the various services, depending on their responsibility (health services, port authorities, hydrological services, police authorities, customs authorities, etc.) Due to the lack of special services for pollution detection and monitoring, this is done by public and private laboratories. The necessary means, i.e. the material for chemical and mechanical dispersion of hydrocarbons are provided by the various enterprises. Sharing of the expenses is not obligatory for municipalities or industries, as the costs are covered by government.

Tanzania and Kenya have no specific law on marine pollution. Protection of bodies of water can be effected through regulations for the disposal of sludge and tailings, sludge channels, poisonous or noxious products resulting from mining operations, and for waste waters. In Kenya, the Port Officer is responsible for the well-being of shipping activities as well as that of the coastal environment. The Port Ordinance prohibits the casting or throwing of any ballast or rubbish and anything likely to form a bank or shoal into ports or upon any shore from which the same is likely to be washed into a port. The Port Officer and the Marine Superintendent exercise the power of determining

61/ Act No. 324, 26 June 1975 on Protection of the Baltic Sea.

whether or not a ship or any other vessel is fit for the transportation of petroleum in bulk.

In an attempt to minimize sea pollution, the Ministry of Public Health of Ethiopia established strict regulations aimed at controlling household waste disposal, and the discharge of industrial wastes comprising oily substances and other dangerous chemicals that may pollute the waters of the sea. The Marine Administration of the Ministry of Transport and Communications issues port regulations which prohibit the discharge of harmful waste materials in the port areas.

Efforts are being made in eastern European countries to control coastal and sea water pollution. 62/ The effectiveness of the measures depends on the legislation in force, the availability of funds and the administrative structure of the relevant authorities. The protection of waters is regarded as an integral part of the general problem of the utilization of water resources. The measures adopted are aimed at preventing further pollution of waters as a result of industrial growth and urbanization, and restoring polluted waters for drinking, household needs, sports and recreation. 63/

In general, the responsibility for the application of water protection measure lies with the organs controlling the factories which discharge waste waters. In the river basin and coastal areas of Yugoslavia, the control of the whole hydrological cycle is given to a single authority in view of the close relationship between fresh and sea waters and of the necessary scientific and research support. Regulation of the conditions and rules governing the discharge of wastes is based on the composition of the water concerned. The degree of purity of water bodies is assessed on the basis of the findings of chemical, bacteriological and biological tests which are different for inland and maritime waters.

Many eastern European countries classify their inland and maritime waters according to usability; sometimes a special status is given to coastal waters, for example in Yugoslavia. The purposes of the classification is to reflect the composition of the water, to show how far waters can be used, to ascertain the causes of pollution, to prepare measures for eliminating pollution, and to establish priorities in the execution of these measures. The indicators for assessing the quality of waters, the system of classification and the methods of analysis are not the same in all countries. The Soviet Union has established purity standards for waters used for sanitary and domestic supplies and for fishing. 64/ Although such standards have been adopted in most eastern European countries, none of the background studies have been developed on a wide scale.

The enterprises and organizations concerned may be penalized financially for any failure to adopt water protection measures. Fines are used for the

62/ The Pollution of Coastal and Estuarial Waters, op. cit., pp. 25, 28, 29 and 31.

63/ New legislation in Bulgaria and Yugoslavia requires the maintenance of the marine environment in its natural state. The discharge of sewage and industrial wastes is not allowed if this natural state is affected, e.g. (Yugoslav) Federal Executive Council Act on Inter-Republic and Inter-State Waters, 10 January 1974.

64/ T. Nagibina, Organization of Water Pollution Control Measures in the U.S.S.R. and Eastern European Countries, WATER POLL./CONF.13, Vol. II, p. 20.

development and rehabilitation of the environment. In some cases, it is more economic for plants to pay fines than build installations to meet the requirements. The suspension or closure of enterprises as a consequence of their violation of protection measures constitutes a grave administrative and legal sanction, and could only be taken in extreme cases. While national councils and individuals have the right in law to claim compensation for damage suffered, or for additional expenses resulting from the pollution and degradation of water, soil, forests and buildings, no serious claims have yet been made.

In Finland, according to the Act of 1962, industries and communities which have over 200 inhabitants must have a permit from the Water Court to discharge waste water into the sea. The procedure for granting permits involves detailed examination. Pollution consequences are examined by a specialized engineer who is appointed by the National Water Board. The terms for the issue of a permit are stated in the Board's report. If the interested party is dissatisfied with the decision of the court, it has the right to appeal to the Supreme Court. The National Water Board offers its experience to the courts that handle pollution cases, controls the amount and the quality of wastes disposed into the sea, and conducts studies on certain industrial wastes. The Institute of Marine Research is an expert body used by the water courts, the coastal communities and industry for the selection of discharge areas. The opinions of this Institute and the National Water Council co-operate in the formulation of a suitable policy for sea protection. Co-operation also exists between the Council of Environmental Protection, the regional planning bodies, the rural community unions and other national and local authorities. In the technical field, it is anticipated that sewer systems which cover 55 per cent of the communities will be expanded to cover 80 per cent during this decade and that biological and chemical treatment of sewage should be completed by 1980. During the same period, the wood-processing industry will complete the change in processes to replace mechanical waste treatment with chemical treatment. It is anticipated that other industries will attain, before 1980, the waste treatment level set for domestic sewage. Investment for the above projects is estimated at one billion Finnish marks. The part to be allocated to the coastal area cannot be estimated separately because part of the wood-processing industry, a large part of other industry and most communities are located on the coast. In addition, it is expected that the load of solid wastes from the wood-processing industry in the whole country will diminish by 80 per cent.

In Norway, pollution aspects are taken into account at the planning stage so as to reduce damage. Norway has established totally-integrated natural resource planning procedures with close administrative co-ordination between water and land use management, and between water management and product control. 65/ The first major legal change was to require all municipalities to prepare overall physical and economic master plans for the use of all resources so as to co-ordinate different sector plans within a community or region. Prior to this requirement, permits were issued by different authorities with responsibility for small sectors of natural resources or environmental management, and there was very little co-ordination. To rectify this shortcoming, the sectoral authorities were brought into the system of

---

65/ For details of this approach, see Instruments for Water Quality Management of Unpolluted and Slightly Polluted Water Resources, E/CONF.70/TP.185, 10 November 1976, a paper submitted by the Government of Norway, Chapter 4 in particular.

master planning on all levels; in the field of water resources for example, it was established at the appropriate level a master plan on the future water use and all major uses of the drainage basin influencing the water quality through pollution. Permits are given only in conformity with the master plan. The Government passed a new Water Pollution Act in 1970 66/ requiring that the approval of water supply and sewage disposal systems (and hence the use of water resources for recipient purposes) as part of the municipal or regional resource plan should incorporate a permit for discharge. This Law also makes it possible to control pollution through legally binding regulations. Such regulations may be combined with permits given on local level, thus decentralizing the permit process. In 1972 a Ministry of Environment was established, to have co-ordinating responsibility for natural resource planning, pollution control and nature conservation. 67/ Under this system, the municipal master plan (general plan) prepared by the community council with public participation is transmitted to the county authorities who make comments and forward it to the Ministry of Environment. The Ministry approves the plan with appropriate suggestions. Both the municipal master plan and the county master plan are advisory in nature. They formulate the pricpal features of land use pattern and naural resource use and development. In this way, they guide the local and county authorities in their local plans for development. The plans become binding once bye-laws have been adopted.

In preparing the master plan, the following points are taken into account:

(a) securing drinking water catchment basins for the future (10 years) either by adopting bye-laws to the master plans that forbid any other use of the catchment area or simply by the county authorities' approval of a municipal plan. For small catchment basins this means no dwellings, no industrial activities, no farming, no motorized transport and so on. For larger catchment areas (i.e. the biggest river in Norway), it may imply rather heavy restrictions on new industrial establishments and changes in production of existing industry. The aim is that all people within a few decades shall drink water practically free from pollution due to human activities.

(b) Areas for residential purposes, industrial development and so on, will be located on the basis of a comprehensive analysis of all important aspects, and great importance will be placed on finding locations where the activities will lead to least harmful environmental effects.

(c) Areas for industry will be zoned so that industry with polluted process water will not be located in drainage areas where such use can harm the planned use of the water resources concerned.

(d) The master plan will show the main future uses and conservation of water resources (rivers, lakes, ground water, fjords) and these uses can, *inter alia*, be made legally binding by adopting appropriate bye-laws to the plan. (Other more direct methods are also being considered).

The Norwegian Government is now in the process of preparing new laws to replace existing planning and pollution laws. The new planning law will stress the co-ordination of municipal, county, and national master planning of all resources. A new comprehensive pollution act will contain regulations on air, water, solid waste and noise to achieve total integrated pollution control by one administrative body.

66/ Act No. 75 of 26 June 1970 relating to Protection against Water Pollution (Water Pollution Act), with amendments of 31 May 1974.
67/ See Chapter 10 (Special Legislative Measures).

As already mentioned, this new Norwegian approach to water resource is part of an overall method for natural resource management. The organizational structure of the authority in charge at the national, county and municipal levels is a very important element. Lack of, or bad co-ordination between the different ministries and between different authorities at the various levels has been recognized as a serious handicap to better management programmes. The Norwegian experience indicates a strong need for co-ordination. It created one ministry having co-ordinating responsibility for approving all physical and economic master planning.

In the light of the above survey, it would appear that most countries seem to have a sufficient body of law regarding marine pollution, even though they may differ in approach and detail. Some of the approaches referred to above, particularly the Norwegian laws, seem to suggest a viable and effective way to tackle the problems in countries having similar situations.

5. Oil Pollution

Oil pollution in the coastal area may come from the different sources of landbased industrial activities and offshore drilling and vessels (discharge or collision). Very few countries have established complete regulations to deal with these sources. National legislation deals mostly with jurisdictional matters and the issuance of authorization to explore and exploit resources. 68/ Some of them contain general provisions for the safety of operations. A number of laws embody more specific reference to the need to prevent pollution during the course of exploration and exploitation of mineral resources of the continental shelf, setting out general rules for oil pollution prevention usually as an adjunct to the regulation of safety. 69/

Generally speaking, those laws which are more comprehensive usually contain such requirements as technical standards of equipment, specific equipment such as blow-out preventers, and regulation of their uses. A number of such laws either prohibit discharges of oil beyond a permissible level (Bahamas, Continental Shelf Act, 1970; Malta, Continental Shelf Act, 1966) or prohibit pollution of the marine environment arising out of such activities in more general terms (Thailand, Petroleum Act, March 1971; Ghana, Minerals (offshore) Regulations 1963 as amended).

Kuwait has a law 70/ which declares all areas within 50 miles from the nearest land to be prohibited zones into which no oil may be discharged. This law, however, does not deal with accidents involving oil spills. It relies heavily on fines as punishment for any contravention. Iran and Bahrain have inserted in their port regulations provisions dealing with oil pollution situations in general. They are concerned mainly with ships' normal operations; they do not deal with oil spills *per se*. Environmental

---

68/ For a detailed analysis of sources, effects, protection measures of pollution in European countries, see *The Pollution of Coastal and Estuarine Waters, op. cit.*

69/ A general review on these questions is contained in "Study of Offshore Mining and Drilling Carried out within the Limits of National Jurisdiction", Working Group of Experts on Environmental Law, Second Session, Geneva, 3-12 April 1978, UNEP/WG.14/2, 23 February 1978; 24 pp.

70/ Kuwait Law No. 12 of 1964 concerning the Prevention of Pollution of Navigable Waters by Oil.

legislation was presented to the Iranian Parliament to establish a 50 mile zone outside its coasts to prevent oil pollution. Oman introduced in 1975 a Marine Pollution Control Law regulating oil discharge from vessels which appears to be the most comprehensive law in the region. 71/

None of the Gulf States has enacted specific legislation covering oil spillage from exploration and exploitation of its continental shelf. There are very general provisions in the offshore oil exploitation agreements. These obligations are usually imposed on operators in the conduct of their activities, e.g. relating to work performance, safety of personnel and equipment, and the obligation to prevent damage as well as oil pollution of the sea. 72/ It has been suggested that legal provisions regarding oil pollution should be strengthened and that oil pollution contingency plans be prepared to mitigate any possible damage. 73/

Kenya is concerned with oil discharges by tankers passing by from the Middle East. Its Petroleum Act provides that specific regulations may be established regarding loading, storage or transhipment of oil. 74/ Rule 48(i) prohibits the discharge of petroleum or water mixed with petroleum from bilges or tanks, of water used for flushing out pipes or of sand used to absorb petroleum. All pipes and connexions on board and on shore between a ship and a quay must be maintained in an efficient condition and they should be free from leakage. In addition, a watch must be constantly maintained near the pipes and connexions during the loading or discharge of petroleum in bulk. Failure to observe the rules relating to the transport of petroleum is an offence punishable by fines. There are however no regulations that are enforceable to tankers passing through the coast.

The Soviet Union adopted in 1953 regulations for the prevention of leakage into reservoirs of oil products, acids and other chemical substances during their transport. 75/ No leakage is permitted when petroleum products and other liquid cargoes are transported on water or to shore. There must be sumps or gutters at the juncture of deck or oil pipes to collect any substance that might seep through. During loading, unloading and transferring of chemicals or liquid cargoes in containers, measures must be taken to prevent such cargoes from leaking into reservoirs. Shipping companies are required to make adequate arrangements for the provision of barges for pumping the residues out of vessels when they are cleaned at ports or wharves. The

---

71/ For example, Oman has a Marine Pollution Control Law (text provided by the Permanent Representative of Oman to the United Nations in a verbal note of 2 December 1974). It deals primarily with discharges from vessels. Israel has an Ordinance to make provision against the discharge or escape of oils into navigable waters.

72/ See M. Hardy, "Offshore Development and Marine Pollution", Ocean Development and International Law Journal, Vol. 1, no. 3, 1973; B.A. Al-Awadhi, Legal Aspects of Maritime Pollution with Particular Reference to the Arabian (Persian) Gulf, Ph.D. thesis, University of London, 1975.

73/ ibid. A number of basic issues are dealt with in Chapter 16.

74/ Similar provisions are also found in Tanzania, the Petroleum Ordinance, Chapter 225, Laws of Tanzania.

75/ These regulations were confirmed by Order of the Ministry of the Maritime and River Fleet of 14 September 1953, No 414.

residue of petroleum products must be pumped into barges or shore storage facilities and brought to the shore and destroyed in places designated by the administration of the port in agreement with the competent authorities.

Belgium recently adopted a plan to combat oil pollution. The plan applies to oil pollution occurring in the territorial sea and on the beach, and is being implemented under the general direction of the Ministry of Internal Affairs. The plan identifies the steps to be taken, defines the respective responsibilities of the various organs involved, and sets out the co-ordination required.

In Portugal, the most serious problem is pollution from hydrocarbons. The examination, measurement and control of pollution is at present pursued under the supervision and co-ordination of the National Association for Scientific Research and Technology and the National Environment Committee. This Committee has an interministerial character and is responsible for the co-ordination of various public services involved with pollution, e.g. Ministries of Health, Public Works, and the Navy.

Measures to prevent marine pollution by oil and oily mixtures from shore installations have been taken by several Mediterranean countries. 76/ Some of the relevant texts deal jointly with oil pollution from ships and from pure installations (the Oil Pollution Regulations in Egypt, the Oil in Navigable Waters Ordinance 1936 (amended) in Israel, and the Port Regulations 1966 in Malta), others are based on the protection of the foreshore (as is the case in Cyprus), while still others include oils among the substances whose discharge into the sea is prohibited or restricted under regulations dealing with waste discharges in general. Among the most specific texts dealing with this problem are a Syrian Law (No. 10 of 26 March 1972), an Order promulgated in Spain on 27 May 1967 prohibiting the discharge by industrial installations of all kinds of petroleum products (or wastes containing such products) into the sea, and a Swiss Ordinance of 19 June 1972 on the protection of waters against pollution by liquids liable to cause a deterioration in water quality.

The Canadian Government has also developed national, regional and local contingency plans, as well as a joint plan with the United States 77/, and is itself taking a very active role in anti-pollution. The arrangements are not dissimilar to those of the United States.

In light of the above experience, a number of points may usefully be mentioned in relation to the role of government in this respect. There should be designated a lead agency or agencies having overall responsibility for preparing and executing plans to combat oil pollution.

If this is not feasible, it may be beneficial to designate a co-ordinating authority; all agencies should be required to co-ordinate their efforts and provide mutual support in preparing for conducting emergency responses.

The Oil in Navigable Waters Act 1971 of the United Kingdom provides powers for government to take action against a damaged ship which threatens pollution to U.K. coasts or waters, comparable to those provided in the 1969 International Convention Relating to Intervention on the High Seas in Cases of Oil Pollution Casualties. These powers extend to action towards foreign ships

---

76/ The State of Marine Pollution in the Mediterranean and Legislative Controls, FAO, Rome 8 September 1972. (mimeographed).

77/ See Manual on Oil Pollution, IMCO, Chapter VII, Section II, 1978.

outside U.K. territorial waters; the primary purpose of this is to remove any doubts about the right under U.K. domestic law to take on the high seas the action envisaged under the Convention. This Act has since been consolidated into the Prevention of Oil Pollution Act 1971.

Under this latter Act, government is responsible for making all the necessary arrangements to deal with oil pollution and to carry out the actual clean-up. It prepares a contingency plan which is based on the probable size of an oil spill and provides the facilities and equipment required according to the plan. In case of a spill, the plan is put into effect and whatever clean-up is required is then carried out by the State. The spiller may be requested to assist the State 78/, but government is in charge of and executes the plan. The assumption is that the State knows best what should be done and will do whatever is necessary for the best interest of the country. Since there is a plan, any necessary action can be taken quickly and immediately without delay.

In the United Kingdom arrangements are in force for reporting vessels which are seen discharging oil; civilian and military aircraft as well as ships of the Royal Navy and Merchant Navy, and fishing and pleasure craft, report such incidents to the Department of Trade via H.M. Coastguard. 79/ If the ships concerned are registered with the United Kingdom or the offence occurs in U.K. territorial waters, prosecution is considered if the evidence is sufficient. If not, the reports are referred to the flag government with a request for investigation, with a view to prosecution. If the ship is scheduled to put into a British port, the U.K. also exercises its right to inspect her records and draw the attention of the country of registry to any evidence which they provide.

The Department of Trade is responsible for the practical measures to deal with oil at sea when it threatens serious coastal pollution, and for drawing up national and district contingency plans. The organizations primarily involved are the Marine Survey Service and H.M. Coastguard. For practical purposes, the coastline is divided into nine districts each with its own Marine Survey Office, which is responsible for initiating sea operations to disperse oil spillages. Local authorities are responsible for clearing any oil which comes ashore.

In the United States, the Federal Water Pollution Control Act, as amended in October 1972, is closely related to the issue of oil pollution. The U.S. Coast Guard's programme in this respect deals with prevention, response and enforcement. Subsection 311(j) (1) (c) requires the issuance of regulations "establishing procedures, methods and equipment, and other requirements for equipment to prevent discharge of oil from vessels and onshore facilities and offshore facilities, and to contain such discharges". The regulations were published in December 1972 after an exchange of views with industry and other interested groups, and following public hearings and review of written comments. The regulations address both vessel operations and terminal transfer operations. Their overall objective is to provide substantial improvement in environmental protection without imposing unreasonable burdens or unattainable standards upon industry.

78/ See P. J. Fallon, "Marine Pollution Control in the United Kingdom", in J.K. Taussig (Ed.), Pollution Control in the Marine Industries, 1974, pp. 297, 318.

79/ "Accidents at Sea Causing Oil Pollution: Review of Contingency Measures", op. cit., p.6.

The Coast Guard has been assigned extensive responsibility and authority to respond to oil pollution in the coastal area. 80/ Surveillance and monitoring by ships, aircraft and shore guards form a major part of Coast Guard efforts to detect discharges. When a discharge occurs, the procedures of the National Oil and Hazardous Substances Pollution Contingency Plan are put into operations. But the Coast Guard must judge the polluter's capability and efforts, and if they are considered inadequate, must take charge and effect the clean-up. 81/ The Coast Guard administers a revolving pollution fund to be used to defray clean-up costs, either when the polluter cannot or will not effect clean-up or when the polluter cannot be identified. The polluter, when known, must with cetain exceptions and within a specified limit of liability, reimburse this fund for the actual costs incurred by the U.S. Government.

The Federal Water Pollution Control Act required each vessel owner calling into a U.S. port to provide evidence of financial responsibility, and also established a $35 m Contingency Fund for use by the Coast Guard in cleaning up spills in U.S. waters. Practically all oil tanker owners and oil companies have entered into voluntary agreements known as TOVALOP and CRISTAL which provide up to $30 m for compensation of oil-spill costs.

The Coast Guard maintains the National Strike Force, capable of responding to an incident anywhere in the coastal United States. Each of the three strike teams is composed of personnel specially trained and equipped for identification of polluters, clean-up operations, and co-ordination of disparate forces, both organized and volunteer. In addition, there are district offices and port captains who are charged with similar duties on a local basis.

Highly-specialized equipment is being developed through Coast Guard research and development efforts to limit quantities discharged, contain the oil on the water, and recover it from the water. Off-the-shelf equipment is stockpiled in areas where effective co-operative clean-up organizations or company-owned resources do not exist. Wide use is also made of qualified contractors, who are specialists in the clean-up business.

The Federal Water Pollution Control Act also stipulates penalty provisions. Failure on the part of a polluter to report a discharge can result in a criminal penalty of up to $10,000 and/or one year in jail. The Coast Guard has authority to assess a civil penalty of up to $5000 for each discharge. The Coast Guard has this civil penalty authority both in the coastal region and in the inland region. Violation of regulations such as those earlier described can bring a civil penalty of as much as $5000.

Other enforcement tools are also available. The Coast Guard can proceed administratively under Revised Statute 4450 against a licensed or certified seaman with a view to revoking or suspending his document. The Coast Guard refers cases of appropriate character to the U.S. Attorney for criminal prosecution under the Refuse Act. One or all of these enforcement tools can be brought to bear, depending upon the facts of the case and the record of the

---

80/ See M. Pitkin, "The Maritime Administration's Pollution Abatement" and S. Wallace, "Coast Guard Pollution Prevention Programme" both in J. K. Taussig (Ed.), Pollution Control in the Marine Industries, 1974, pp. 371-378 and 165-176 respectively.

81/ See Multi-Agency Oil and Hazardous Materials Pollution Contingency Plan: New York, New Jersey, Puerto Rico and Virgin Islands, published by U.S. Coast Guard, 2 volumes, Washington, D.C., 1976.

violator. The Pollution Control Act also requires administrative action where a discharge occurs and the polluter is identified; a civil penalty must be assessed. Only the factors specified by the statute may be taken into account in determining the amount of the penalty. The Coast Guard does not have the discretion to limit the penalty to a letter of warning. It is not allowed to take into account the actions of the polluter after the discharge occurs, i.e. the promptness and efficiency of the clean-up effort. The purpose is to stress the importance of penalty.

The United States and Canada concluded in 1973 a joint Contingency Plan to control oil spills. It may be useful to mention some of its main features. Its objectives are to develop preparedness for the discovery and reporting of a pollution incident, and the prompt execution of joint measures to respond to such an incident. 82/ The Plan outlines standard alerting and notification procedures, command structures, clean-up procedures, post clean-up requirements, and arrangements for assuming the responsibility for the cost of operations. The Plan also makes arrangements for the organization of the input of all responsible agencies and authorities in both countries. The on-scene commanders and deputy commanders are designated beforehand. They co-ordinate manpower and resources, and control the clean-up activities with the advice and support of the Joint Response Teams, which consist of representatives of directly-involved federal, provincial and state agencies of the two countries. One of the most positive features of the Plan is the opportunity it affords each party to utilize the full resources of the other for a severe incident confined within its own boundaries.

Under the Plan, the response to an oil pollution incident is separated into four distinct phases:

(i) Discovery and alarm. Border communications between all levels of government are recognized as vital to the success of the Plan. Any pollution incident which presents a threat to the other party must be promptly reported to the appropriate agency of that party in accordance with the Plan.

(ii) Evaluation and plan invocation. When a situation falls within the scope of the Plan, it is required that designated agencies take advantage of any and all available resources and arrangements provided for in the annexes to the Plan in order to mobilize regional forces. The on-scene commander will issue an initial report to the Joint Response Team. The Chairman of the Team can invoke the Plan if the situation warrants it. The Joint Response Team is an administrative and advisory component and is not a field-operating force. It makes recommendations to the on-scene commander, handles public information during incidents, keeps authorities at the national level appraised of the situation, and reviews contingency planning from time to time to detect any weaknesses which may have developed and may need be corrected, reporting these to the appropriate authorities.

(iii) Containment and countermeasures. When the Plan is invoked, the on-scene commander of the control and clean-up activity is to be represented by the country most seriously endangered. If, as containment efforts continue, the severity of the threat shifts, the on-scene commander is changed.

82/ See G. H. Brown, "The Joint Canada-United States Marine Pollution Contingency Plan for Spills of Oil and other Noxious Substances", in T. F. P. Sullivan (Ed.), Pollution Control in the Marine Industries, International Association for Pollution Control, 1973, pp. 68-81.

(iv) Clean-up and disposal. The on-scene commander establishes the clean-up strategy and has the final word on the deployment of available manpower and equipment.

The cost of operations of both parties is to be borne by the party responsible under the Plan for the waters in which the pollution occurred. Special arrangements have been made by each country for customs and immigration clearances for response personnel and equipment.

It would appear that not many countries have oil pollution contingency plans, and that such plans would be important for coastal protection particularly in those areas where large quantities of oil are involved.

CHAPTER 12

Natural Resource Management Approach

Several countries have chosen natural resources as the central subject for management irrespective of their location. While this is a thematic approach, it has yielded certain benefits to the coastal area. The measures taken in Venezuela, Jamaica and Norway are referred to in subsection 1 as example.

Subsections 2 and 3 contain information on the protection and development policies regarding living resources, and information relating to mineral resources and industrial development. Their purpose is to illustrate the issues connected with the development of natural resources in coastal areas.

1. Examples of Natural Resource Approach

By the Decree of 1974, 1/ Venezuela established a Commission to undertake an evaluation of all the problems concerned with the use of renewable natural resources in the islands, lagoons and coasts. Its members are composed of representatives from the Ministries of Agriculture and Ranching, of Foreign Affairs, of Defence, and of Health and Public Welfare.

The specific duties of the Commission are (i) to effect an evaluation of all the problems that confront the renewable resources of the coastal zone; (ii) to undertake a census of the existing population within the coastal area with an indication of ownership; (iii) to elaborate a legal, sociological, sanitary and economic framework, with the recommendations necessary for the solution of the diverse problems that exist in the coastal area; (iv) to propose wildlife refuges in coastal areas which will serve as protected zones and recreation areas; (v) to propose measures preventing activities which may damage coastal ecology. Thus, there is a basis for taking the necessary action in the coastal areas. Consequently, the Commission undertook a census of projected development in population and property, and is currently compiling data relating to legal, sociological, sanitary and economic considerations in different areas in order to recommend appropriate action.

The Commission has also identified a wide variety of problems and issues in the coastal area and has recommended to the President the establishment of a permanent government body of a unique and definitive nature for the preservation of the environment with special attention to the coastal zone of

---

1/ Decree 122 of 27 May 1974.

Venezuela. 2/ In 1977, a Ministry for the Environment and Renewable Natural Resources was established to continue all the activities assumed by the Commission.

In 1972. the Los Roques archipelago was designated as a national park to preserve that area, and decrees were issued to outlaw the filling of coral structures and mangrove swamps on the coast of the mainland and on the islands in the Caribbean Sea. Various quays have been built on the waters of gulfs, bays, islands, and lakes in the vicinity of Chichiriviche, Morrocoy, and Tucaces; these existing constructions must be demolished within a specified period. Coral reefs and mangrove trees may not be damaged or destroyed. Untreated waste waters may not be discharged into the region; likewise dredging and underwater fishing are forbidden. The size and speed of boats navigating in the waters are limited. The Ministry of Agriculture and Ranching, in co-ordination with the Ministries of Interior, Development, Defence and Health, is to enforce the Law. A 500m protective zone has already been drawn around lakes and lagoons. Within this area, no farming or ranching and no destruction of vegetation is permitted without government consent. The Director of Renewable Natural Resources is to supervise this scheme. In addition, a series of decrees has been issued for the above-mentioned purposes. 3/

A National Parks Institute was created as an independent agency to administer the parks system. 4/ Responsible for planning, conservation, and education, the Institute is required to issue rules and regulations for ensuring the health and safety of persons in the parks, as well as for preserving the natural resources. Reserves, refuges and sanctuaries may also be established to protect mammals, birds, reptiles and amphibians. 5/ Hunting is prohibited both in the refuges and sanctuaries; threatened species may be transferred to sanctuaries for special protection to increase their population. Reserves are zones set aside for orderly development of the wildlife population and for hunting under controlled conditions. In the coastal zone, Venezuela has used this authority to declare certain quays and islands near the State of Falcon a wildlife refuge. This action was taken primarily because a number of migratory birds have their nesting sites in this area. 6/

The filling of coral structures and mangrove swamps on the coast of the mainland and on the islands in the Caribbean Sea has been prohibited in order to prevent ecological changes threatening coral formations and mangrove trees. Houses may not be built over the water in or near the coast or the islands of that Sea. The Law also prohibits the cutting or destruction of mangrove trees, as well as the dredging of canals or of marine bottoms in mangrove swamps. Nor can polluted water be discharged into these swamps. Artificial islands may be installed in these waters only with prior permission of the Ministry of Health and Public Welfare and the Ministry of Agriculture and Ranching; these permits are in addition to those already required under

2/ Informe para el Decreto No. 112 del 26-5-74.

3/ Regulation of Ministry of Agriculture and Ranching, 4 July 1974.

4/ Law of National Park Institute, 15 October 1973.

5/ Wildlife Protection Law, 11 August 1970.

6/ Law 911, 2 June 1972.

the Law of Navigation. A recent regulation orders the demolition of "palafitos" 7/ 8/ and other structures built on the waters among the mangrove trees in the Bay of Buche. The Ministries of Interior, of Defence, of Health and Public Welfare, and of Agriculture and Ranching are to co-ordinate and execute the regulations. Another decree prohibits the construction of houses, cabins, or permanent camp grounds on the shores of Caicaro Lagoon. 9/ This lagoon is now reserved for sports fishing only and the size of permitted boats is limited.

There is a basic framework in Venezuela for tourism development in the coastal area. Under the Law of Tourism, the National Tourism Corporation may request the President of the nation to declare any region, place or building as being " of public utility", when desirable for the development of tourism.10/ The Corporation has been given jurisdiction over land and other areas on the shores of the oceans, rivers, lakes and lagoons in a zone extending from the low-water line to 500 m inland. The Corporation is responsible for the inspection, maintenance, order, and conservation of these areas. No buildings may be constructed, works undertaken or services rendered in this zone without prior approval of the Corporation. The Corporation may also purchase land or personal property for the purpose of tourism development. Only companies, organizations and persons registered with the Corporation may operate in the tourism industry. Violations of the laws or of the Corporation's rulings are punishable by fines or by revocation of tourism industry permits; the Ministry of Development, upon request of the Corporation, enforces these penalties.

The Paraguana Peninsula, the San Luis Beach, and Tortuga Island have all been declared as being "of public utility and touristic interest". 11/ The shores of the "Capital Region" have also been so classified; the Corporation is charged with designing a plan for the best development of this area from a touristic and recreational standpoint. The area where the Copey Channel meets the Caribbean Sea has likewise been declared, and both the Corporation and the State of Miranda are empowered to buy or expropriate land within this region 12/ The Corporation is to plan the construction of tourist and recreational facilities in the zone, and the plan is to be executed by the Ministries of Public Works, of Development, and of Foreign Relations.

In Jamaica, the traditional lack of a co-ordinated approach towards the management of the island's natural resources is changing towards a recognition of the need for the integrated and co-ordinated functions of agencies

---

7/ Decree 110 of 26 May 1974.

8/ Regulation of 18 November 1974. Although the dictionary meaning of "palafitos" covers only "primitive dwellings", in actual fact all illegal structures, including luxurious houses, are to be demolished under this Regulation.

9/ Regulation of Ministry of Agriculture and Ranching, 4 July 1974.

10/ Law of Tourism 22 July 1973 and Decree 1673 of 9 March 1974.

11/ Decree 456 of 3 October 1974; Decree 144 of 5 June 1974; Decree 1675 of 9 March 1974.

12/ Decree 251 of 31 July 1974

involved in the management of the environment. As part of the strategy for conservation of the natural resources, including coastal resources, the Government of Jamaica established the Natural Resources Conservation Authority in 1975 by amalgamating the following agencies: the Beach Control Authority; the Watersheds Protection Commission; the Wildlife Protection Committee; the Natural Resources Planning Unit; the Marine Advisory Committee; and the Kingston Harbour Quality Monitoring Committee.

The Authority currently operates under the Beach Control Act, the Wildlife Act; the Watersheds Act and the Flood Control Act.

Similar to Venezuela's Committee on evaluation of all problems confronting the renewable resources in its islands, lagoons and coasts, Norway has established a totally integrated natural resource planning procedure with close administrative co-ordination between water and land use management and between water management and product control. 13/ The legislative changes have been introduced to require all municipalities to prepare overall physical and economic master plans on the use of all resources so as to co-ordinate different sectoral plans within a community or region. Prior to this change, development permits in the coastal areas, as in other areas, were issued by different authorities with responsibility for small sectors of natural resources or environmental management. There was very little co-ordination. To rectify this shortcoming, the sectoral authorities have been brought into the system of master planning on all levels. In the field of natural and water resource planning, for example, there has been established at the appropriate level a master plan on natural resources, and on the future water use and all major uses of the drainage basin influencing the quality of water through pollution. Development permits are then given only in conformity with the master plan.

In 1972, a Ministry of Environment was established to have co-ordinating responsibility for natural resource planning, pollution control and nature conservation. 14/ Under this system, the municipal master plan (general plan) prepared by the community council with public participation has to be transmitted to the county authorities for comments. The county also prepares a master plan with the same procedure and then submits it to the Ministry of Environment. Plans are harmonized at each level. Both the municipal master plan and the county master plan are advisory in nature. They formulate the principal features of land use pattern and natural resource use and development. In this way, they guide the authorities in their local plans for development, since guidelines are given for preparing master plans; the plans become binding once bye-laws have been adopted.

The Norwegian Government is now in the process of preparing new laws to replace existing planning laws. The new planning law will stress the co-ordination of municipal, county, and national master planning of all resources.

This new Norwegian approach to natural resources emphasizes the importance of the organizational structure of the authority in charge at the national, county and municipal levels. Lack of, or bad co-ordination between different

---

13/ For details of this approach, see Instruments for Water Quality Management of Unpolluted and Slightly Polluted Water Resources, E/CONF.70/TP.185, 10 November 1976, a paper submitted by the Government of Norway, Chapter 4 in particular.

14/ See Chapter 10 (Special Legislative Measures).

ministries and between different authorities at the various levels has been recognized as a serious handicap to better management programmes. The Norwegian experience indicated a strong need for co-ordination, and it created one ministry having co-ordinating responsibility for approving all physical and economic master planning.

2. Protection and Development Policies Regarding Living Resources

Almost all Latin American countries have established some protection and development policies regarding living resources. In general, a licence is required for commercial and in most cases, sport fishing by national or non-national 15/ in waters under their jurisdiction. Various rules have been established to exclude non-nationals from fishing activities or to limit them, particularly in certain zones reserved for nationals. The legislation of Brazil, Mexico and Peru illustrates this approach.

Brazil permits certain foreign vessels to conduct fishing activities in the outer 100 mile zone of her territorial sea. 16/ Even within the outer zone, the fishing of crustaceans and "other living resources which are closely dependent on the seabed under the territorial sea", e.g. lobster, is reserved for Brazilian vessels only. In the inner zone, foreign vessels may be permitted only if compensation is paid. The 1972 Agreement with the United States provided that the latter agreed to license and control the number of American shrimp boats operating in Brazil's 200 mile zone. The Agreement further authorizes Brazilian officials to board and search American vessels in that zone and to seize and detain any vessel reasonably believed to be in violation of that Agreement. Proceedings against such ships are reserved to the United States, and Brazil has to hand over the detained vessels for prosecution in the United States. Certain payments are to be made to Brazil to cover its share of the cost of enforcing the Agreement. A recent Brazilian statute further clarifies this situation by stating that foreign vessels can fish in her territorial waters only with a permit from the Ministry of Agriculture or when authorized to fish pursuant to an international accord. 17/

Subject to certain exceptions, commercial fishing by foreign vessels within the 200 mile Mexican economic zone is prohibited. 18/ The executive branch is to determine the allowable catches of the living resources in this zone. When the total allowable catch is greater than the hunting and fishing capacity of Mexican vessels, the Executive can grant foreign vessels access to the surplus of the allowable catch. Foreigners wishing to obtain such permits must register and must agree to certain specific conditions, such as not to unload their catch within Mexican territory, to make available to Mexican nationals on a gratuitous basis the technology used in the fishing operations and in the processing of authorized catches, and to make a cash deposit as a guarantee of compliance. In addition, at least 50 per cent of the crew must be Mexican nationals hired in that territory, and there must be equal pay for both foreign and Mexican crew. Foreign vessels are not allowed to catch anchoveta

---

15/ For example, Venezuela Fishing Law of 1944 (Article 13); Peru, Decreto Ley 18810 of 1971, (Articles 6 and 7); Argentine, Buenos Aires, Decree 2269, 4 May 1965 (Articles 3 and 10) and Law 17500, 25 October 1967 (Articles 1 and 2); Brazil, Decree 68459, 1 April 1971.

16/ Decreto Ley 1098, 25 March 1970.

17/ Law 6276, 1 December 1975.

18/ Law on the Exclusive Economic Zones, Decreto Oficial, 13 February 1976.

or sardines. The basic objective of this Law is to reserve future potential development for Mexican nationals. 19/

Peru has special regulations designed to protect its national interest in fisheries. Thus, no foreign participation in fishing companies may exceed 49 per cent. No fishing permit can be granted to enterprises with foreign participation whose main objective is to manufacture fish meal or oil for indirect human consumption. Every private fishing enterprise in Peru is required by law to ensure that the workers will receive 20 per cent of the net profit and that this sum will normally be reinvested in the company's own shares. This scheme is designed to give the workers a voice in the management of the firm and ultimately a significant part of the equity ownership of the company. Foreign vessels may be permitted to fish within Peru's 200 mile zone if they are properly registered with and licensed by the Ministry of Fisheries. The registration fee is $500 per ship and a licence fee by tonnage is imposed. 20/ Restrictions are placed on the right of such foreign vessels to sell their catches in the Peruvian market.

In view of the importance of fisheries, several Latin American countries have created special agencies to deal with fisheries. Venezuela, for example, has a National Consultative Council on Fishing Policy, whose members include both representatives from government and research institutes. The Council makes recommendations on the development and financing of the fishing industry, as well as on the enforcement of fishing regulations. Peru has established a separate Ministry of Fishing. 21/ Ecuador has estblished a General Office for Marine Development, which co-ordinates the activities of the Merchant Marine, the Army Oceanographic Institute and the Naval Transport Company. 22/

Usually, government agencies authorized to grant licences also have power to set fishing seasons for different species, to limit the size of fish which may be taken, and prohibit the use of certain kinds of gear. 23/ Generally speaking, harpoons or dredgers, powder explosives, lime, sulphur, acids or other chemicals which may injure aquatic fauna are prohibited. In Venezuela, fishing areas are determined according to the size of fishing boat, e.g. boats larger than 18 m are not permitted to fish within 6 miles of the coastline. The purpose of this is to reserve nearshore fishing for smaller fisherman with

---

19/ It is reported that so far, Mexico has exploited only about 10 per cent of her huge coast. 98 per cent of the scientific research on resources adjacent to her coast has been carried out by foreign powers without Mexican participation. Enforcement of these new provisions, however, may prove difficult, since as of 1973 Mexico had only 69 units to police her entire coastline.

20/ Supreme Decree 011-71-PE-1971.

21/ Decreto Ley 18026 of 1969.

22/ Decree 62 of 10 June 1969.

23/ Venezuela, Regulation of Ministry of Agriculture and Ranching, 6 July 1974; Argentina, Decreto Ley 2269, 4 May 1965; Brazil, Decreto Ley 454, 5 February 1969.

simple equipment. The Ministry of Agriculture and Ranching may designate certain species to be used only for fish fertilizer, flour or oil, and it is then illegal to use other species for these purposes. 24/

In some parts of Argentina, a licence is required for extracting rushes or weeds from its waters; 25/ in other parts, the extraction or harvesting of algae from the beaches or sea in a three-mile zone extending outward from the coastline requires special permission. 26/ The provisional government is empowered to issue regulations to protect and conserve algae, e.g. seasons and areas for harvesting, as well as to form companies eligible to collect algae.

There are also laws designed to protect specific species of marine life. For example, Brazil and Venezuela are parties to the 1966 International Convention on the Conservation of Atlantic Tuna, 27/ which established thereunder an International Commission. The Commission is empowered to issue recommendations designed to maintain the population of tuna and tuna-like fish. Such recommendations become binding on the member states within sixty days, if no objection is raised. Mexico as a party to the Convention for the Regulation of Whaling, is bound by Regulations of the International Whaling Commission. This body may, for example, set seasos for whale hunting and designate protected species. 28/ A 1973 Mexican regulation established a protected area for whales on the Pacific coast. 29/

The taking of turtles or turtle eggs along the coasts or islands of the Orinoco River in Venezuela has been prohibited since 1972 for a five-year period. 30/ To further protect these animals, camping and the embarking or disembarking of boats on certain islands in this river, except in case of emergency, is forbidden. In the Caribbean Sea, a green turtle (chelonia mydas) refuge was declared in 1972, emcompassing the Aves Islands, the corresponding continental shelf and the territorial sea. 31/ The Ministry of Agriculture and Ranching is charged with enforcing these laws. In 1975, Peru suspended cod fishing around the area of Port Haurmey.

Lobster fishing has been forbidden in Venezuela since 1975. An exception to this prohibition was reserved for the resident fishermen of the Los Roques archipelago. 32/ Local fishermen in that area must obtain permits from and register each trap with the National Fishing Office. The total number of permits issued in any one season is limited, and the commercial sale or transport of lobsters is also regulated.

24/ Fishing Law 6 October 1944. Peru established PESCA-PERU, a corporation owned wholly by the Government, to produce fish meal and oil.

25/ Decree 2269 of 4 May 1965, Article 45.

26/ Chubut Law 939, 23 July 1972.

27/ Text of the Treaty in 673 UNTS p. 63.

28/ Text of the Treaty 161 UNTS p. 72.

29/ Diario Oficial, 14 January 1972.

30/ Regulation of Ministry of Agriculture and Rachning, 20 May 1972.

31/ Decree 1069, 23 August 1972.

32/ Regulation of Ministry of Agriculture and Ranching, 4 July 1975.

Brazil limits fishing of pink shrimp in the open sea between the 20th and 30th latitudes to vessels in excess of five tons. Such vessels must be licensed by the National Fishing Agency. Exceptions are made for shrimp boats under construction or approved for construction prior to its effective date. The avowed purpose of this legislation is to foster increased production of large shrimp boats.

Pearl gathering and oyster fishing in Venezuela are subject to special laws which establish fishing seasons, and prohibits taking oysters under certain sizes. 33/ Peru makes trading in pearls a government monopoly to be administered by the Peruvian Mineral Bank. 34/ The capture of ornamental fish is also regulated.

3. Mineral Resource and Industrial Development

In both Kenya and Tanzania, ownership of mines in coastal areas is vested in the government or in the Head of State who holds such ownership in trust for or on behalf of the government. 35/ This state ownership of mineral resources stems from a cardinal principle that all the country's natural resources are the property of the nation and that the government acting on its behalf should ensure that their exploitation benefits the whole nation.

The powers and functions of the agencies and institutions charged with overseeing mining and related activities in coastal areas are the same in Kenya and Tanzania. In the former case, for example, the mining industry is governed by the Mining Act which deals primarily with precious metals and stones. The extraction of non-precious minerals, e.g. clay, sand and gravel, is not governed by the Act and remains unregulated to a large extent. Prospecting of minerals requires authorization from the government and cannot take place in (i) land dedicated or set aside for burial or for public purpose, e.g. parks; (ii) the foreshore between high and low-water marks at ordinary spring tides within the boundaries of any municipalities or urban centres or (iii) any area, which is the site of or within a certain distance, e.g. 50 or 100 m, from any dam, canal, reservoir or building belonging to the government or a county council. It is specifically provided in the Tanzanian legislation that mining cannot take place in any national park, forest reserve, or game reserve.

The cutting down of timber or the clearing away of vegetation in coastal area is permitted for the construction of works and the installation of machinery connected with mining. The cutting and making use of any trees in the course of mining operations are subject to payment of fees or royalties under the law relating to forests. Where mining operations are carried out near bodies of water or may pollute or interfere with the flow of such bodies of water, the Minister of Natural Resources is empowered to make regulations

---

33/ Regulation of Ministry of Agriculture and Ranching, 4 July 1974.

34/ Decreto Ley 18882 of 1971.

35/ At present, mineral resource development in Kenya and Tanzania is still relatively small. Kaolin is found in the coastal area of Tanzania (the Pugu Hills) and is used in the chemical and ceramic industries of Dar es Salaam. The Uluguru Mountains are the main mica-producing area in the country. Copper and manganese deposits are found in the coastal regions near Mombassa (Kenya), at Vitengeni and Mrima Hill. Drilling of oil near the coast inland from Lamu is also taking place. Since large coastal areas in both countries have not yet been surveyed, whether other minerals deposits exist is not known.

to protect such bodies of water against activities which may damage the quality or usefulness of such water. 36/ For example, Regulation 136(2) provides that no water containing any poisonous or injurious substances be allowed to find its way into any stream, lake, dam or reservoirs without having been first rendered harmless. On abandoning mining operations, the location holder is required to fill in pits, holes and excavations resulting from his mining activities. 37/ The regulation does not specify any particular methods for filling in such pits and excavations, the only requirement being that excavations be filled in or otherwise rendered safe to the satisfaction of the Commissioner of Mines and Geology.

Control over the mining operations is maintained by the Commissioner who ensures that the holder of a location complies with the terms and conditions upon which the location was granted; he also ensures that the holder does not mine any mineral other than the one for which the location was granted. It is also his duty to oversee the restoration of land and to adjudicate any disputes which may arise from the activities. His decree or order is enforcible by the court having the jurisdiction.

In Kenya, the responsibility for survey and mapping of mineral resources and for the control of mining activities is assigned to the Mines and Geological Department of the Ministry of Natural Resources. An Economic Investigation Programme has now been set up under the Department to undertake research and evaluation work on ore minerals, building materials, mineral procession and mineral economics.

Kenya does not allow an oil mining lease more than 1250 km^2. While an oil prospecting or oil exploration licence can be granted to any company, an oil mining lease can only be granted to a company incorporated in Kenya or a company incorporated outside but registered in Kenya.

There is no specific reference to protection of the environment. The only recourse which may serve as a preventive measure is the provision requiring compensation for causing "disturbance" to the owner or occupier of private land or causing damage to such land or to any crops, trees, buildings, stock or works thereon. A person who is neither an owner nor occupier of the land in question does not seem to have the right to seek compensation for disturbance or damage.

In most Latin American countries, laws in the fields of mineral resource extraction, like those in fisheries, are mainly directed towards excluding or limiting exploitation by non-nationals. Mineral deposits are generally designated as property of the State. 38/ The Mexican Constitution, for example, specifically claims public ownership of the mineral deposits in the 200 miles exclusive economic zone. Concessions may be granted, however, for

---

36/ Legal Notice No. 275 of 1969.

37/ According to the Mining Law of the German Democratic Republic, a miner is responsible for reclaiming land for agriculture and forestry purposes after mining has been completed. In this manner, about 2000 acres of land have been rehabilitated, among which 40 per cent has been used for agriculture and 60 per cent for forestry.

38/ For example, Peru (Decreto Ley 18880 of 1971, Art. 1) Mexico (Constitution, 1961, Art. 136), Venezuela (Constitution, Art. 27). Ecuador also declared the "inalienable interest" of the State in "petroleum exploration and conservation" (Decree of 6 June 1972).

exploitation. Upon expiration of the time period of a mining concession, such property reverts to the State.

Until 1974, the exploration of salt in Venezuela was done under licences and contracts. 39/ Since 1974, the National Salt Company (ENSAL) has been authorized to use and to purchase from concessionaires all the installations, equipment and property necessary to continue the mining and processing of salt. The expressed purpose was to ensure the proper iodinization and refining of salt for public health reasons. Nevertheless, ENSAL may contract with mixed capital companies 40/ to refine and iodize the salt. A commission was created to oversee the operations of ENSAL, and its members consist of representatives of the Ministries of Health and Public Welfare, the Treasury and of Development. This commission must approve both the contracts by ENSAL to buy the properties of the former concessionaires and the contracts of ENSAL with other companies to process salt.

Peru has created a wholly state-owned corporation, the Salt Company, whose capital is furnished by the State; its board of directors consists of representatives from the Ministries of Industry and Commerce, and Energy and Mines, and of the Salt Company's workers. 41/ Mineral wastes must be disposed of in special dumping grounds where proper safety measures are to be taken to prevent contamination of water and soil. The Ministry of Mining, upon the advice of other appropriate agencies, is to establish such special dumping grounds and authorize the installation of mineral compacting plants.

The Yugoslavia Basic Law on Mining Ores of 18 February 1966 also applies to the extraction and exploitation of mineral ores on the seabed, underneath it, or outside the coastal sea of Yugoslavia but within its national jurisdiction. In addition, the Law of 22 May 1965 made special provisions for exploration and exploitation of the continental shelf. Such activities must not unjustifiably interfere with navigation, fishing, protection of living resources of the sea in general and with fundamental oceanographic or other scientific research.

In South Asia, mining has always been considered a concern of the State. In India and Pakistan, for example, central government has the sole responsibility for nuclear substances, oilfields and gasfields; other mines are in the hands of the provincial government. 42/ For any group or class of mines, there is a Mining Board, and the development of mines is in general in the hands of corporations set up specifically for that purpose. The government plays an important role in all industrial development both in terms of organization and financing. Prospecting or mining operations are under licence or lease and usually only nationals may hold such licences or leases. Leases are limited in the first instance to 30 years for coal, iron ore, and 20 years for any

---

39/ According to Constitution, Art. 136 (10) and Decree 560, 26 November 1974, salt deposits may not be alienated.

40/ A mixed capital company means a corporation, part of whose stocks is held by the government and part by private persons or enterprises.

41/ Law 17525 implemented by Decreto Ley 18923 of 1971 and 18350 of 1971.

42/ For example, the Mines Act, 35 of 1952 of India as amended by 52 of 1970.

other mineral. Mines are government controlled, 43/ and the rules for the granting of prospecting licences and mining leases and for the conservation and development of minerals are also made by central government. The Geological Survey may be authorized to make surveys. 44/

The import, transport, storage, production, refining and blending of petroleum and other inflammable substances are also controlled by government.45/ In India, for example, no one is permitted to produce, refine, or blend petroleum save in accordance with rules made by central government. Pipelines must not be laid under residential land, permanent structures, lands close to a dwelling house, or less than 1 m from the surface. 46/ Where pipelines have been laid, no building, excavation or construction of tanks, well or reservoirs can be made without special authorization. In Sri Lanka, the local authorities have powers to license the import, transport and storage of oil, and the minister determines ports of entry.

The development of oilfields is governed by special legislation in India.47/ Central government makes the rules for granting leases and for the conservation and development of mineral oils. There is an Oil and Natural Gas Commission whose functions are to plan, promote, organize and implement programmes for development of petroleum resources; it makes surveys and undertakes exploration (including drilling) and geological and other investigations. The Commission may set up an industry with government approval. It has power to acquire land for that purpose and to make rules and regulations with the prior approval of central government, to discharge its functions and regulate its own procedure, central government having the power to amend or rescind any regulation. A similar course is taken in Pakistan. 48/

In South Asia, coastal regions are favourite locations for factories as they are near to sources of power, points of entry and departure. They also provide access to cheap labour forces. There are laws dealing with various aspects of factories, such as licensing and registration, inspection, health safety, welfare, working hours, employment and young persons, and annual leaves. 49/ The question of discharges from factories into rivers or the sea

---

43/ See For example, (India) Mines and Minerals Regulation and Development Act, 67 of 1957.

44/ The amending Act, 56 of 1972.

45/ See for example, the India Petroleum Act, 30 of 1934, as amended by 24 of 1970. The Act is also in force in Pakistan and Burma.

46/ The (India) Petroleum Pipelines (Acquisition of Right of User of Land) Act, 50 of 1962.

47/ The Oilfields (Regulation and Development) Act, 53 of 1948, as amended by 39 of 1969.

48/ The (Pakistan) Regulation of Mines and Oilfields and Mineral Development (Government Control) Act, 24 of 1948, as amended by Ordinance 8 of 1971; the Oil and Gas Development Corporation Ordinance, 37 of 1961 as amended by Act 15 of 1966 and 3 of 1971.

49/ Some examples include: (India) Factories Act 63 of 1948; (Pakistan) Older Factories Act, 25 of 1934, as amended Ordinance 6 of 1966; (Sri Lanka) Cap 128 of the Laws of Ceylon.

generally figures of little importance in the legislation. Some general guidelines dealing with marine or coastal pollution resulting from land-based sources appear to be needed. Legislation exists in Pakistan and Burma pertaining to scientific and industrial research.50/ In Sri Lanka, an Institute of Scientific and Industrial Research was set up under the control of a minister to develop natural resources and industries. 51/

In South Asia, the government plays an important role in all industrial development both in terms of organization and financing. In India, 52/ for example, the development of any industries requiring finance comes under the State Financial Corporation Act (63 of 1951) whereby a corporation is set up to guarantee loans from banks and deferred payments to underwrite stocks, and to grant loans and subsidies against debentures to "industrial concerns". The Act is applicable to all major industries, including processing of goods, mining, hotels, transport, generation of power or development of any land contiguous to an industry area. This Act thus provides a financial framework for the creation of new industries. Central government may establish a Development Council for any new industry. Existing industries are registered and new undertakings are licensed by central government, which also has the power to investigate industrial projects and issue directives for undertakings. Direct management and control of industrial undertakings may be assumed by central government, and it may control supply, distribution and prices.

An interesting code is the West Bengal Industrial Infrastructure Development Corporation Act (25 of 1974) which established a Corporation to secure the development of trade, commerce and industries on a well-planned basis. Its functions include reclamation and improvement of land, acquisition of land, sewerage, transport, roads and zoning. The Corporation's powers of land acquisition are provided in detail. No person may undertake any specific development of land in any industrial estate save with the permission and under conditions laid down by the Corporation. This Act appears to be a typical legal framework in South Asia. It provides for the tight and exhaustive hold over the Corporation by the State Government and the relatively loose and flexible and vaguely-expressed delimitation of the Corporation's own powers vis-à-vis the public and its actual duties. The legal framework is widely drawn to enable the Corporation to equip itself to function multifariously. The Act sets up a framework for action. Whether the action itself takes place depends on the interaction of the designated officials of government and their contacts in the enterprises in question.

In Pakistan 53/ some industrial development can only take place under the control of central government (arms, nuclear materials, pétroleum and mines), while the rest comes under provincial control. The appropriate government may

---

50/ The Pakistan Science Foundation Act, 3 of 1973, set up a Financing agency; the (Burma) Union of Burma Applied Research Institute Act 1954, set up an Institute to further research.

51/ Cap. 164 (Sri Lanka), as amended by 14 of 1962 and the Law of 1973 (No. 51). There is also the National Science Council of Sri Lanka under Law 36 of 1975.

52/ The Industries (Development and Regulation) Act, 65 of 1951.

53/ The (Pakistan) Development of Industries (Government Control) Act, 13 of 1949.

plan and regulate the setting-up of new undertakings and the development of existing undertakings and make rules to carry this out. Industries such as oil, gas and electricity are under federal control and are not governed by the law of companies. 54/ There is a Board of Industrial Management whose function is to ensure that establishments are managed efficiently and in accordance with sound business principles. The Board may also give directions to managing directors in the exercise of their powers. There is a plethora of statute law in the provinces of Pakistan, dealing with further development of cottage and small industries. Loans and guarantees may be provided for surveys, and for establishing facility centres, small industries trading estates and artisan colonies.

Many developing countries have special legislation regarding production of electricity. For example, in Kenya, electricity-generating activities are controlled by government through a licensing system. 55/ The applicant for a licence to generate energy must submit an application and make public the application. Any person, company, public or local authority who desires to make representation or objection regarding the application is given an opportunity to do so and his representation or objection must be made by letter addressed to the minister in charge, with the same procedure being followed where it is desired to have any restrictions to be inserted in the licence. The application, together with any representation or objection made, is considered after the expiry of sixty days and the minister in charge may then refuse the application or grant it on such terms or conditions as he may impose.

In the event the licensee cannot obtain land on his own for the purposes of constructing a generating station, the government may facilitate the acquisition of the land by authorizing compulsory acquisition. Some protection measures for the environment are provided by Section 69(1) of the Electric Power Act which imposes liability on all licensees, operators or owners for nuisance caused or permitted by them. Specific protection is extended to shores, seabed, canals, creeks, bays and estuaries, whereby no licensee may have the right to take, use or interfere with any portion of such areas. In effect, the construction of generating stations on the country's shores, seabeds, creeks and bays is barred since such construction would invariably involve the use of a portion(s) of the said shores, seabeds, creeks and bays.

In Tanzania, one of the main sources of hydroelectricity is the Grand Pangani Falls, about 40 miles from the coast on the Pangani River. Unlike the Kenyan system, it is the President who has the power to grant a licence for the supply of electricity and to expropriate land for that purpose. There is a general ordinance dealing with electricity. The Electricity Ordinance empowers the Minister to appoint elctricity inspectors who supervise the construction of works and may enter any premises of a licensee to ascertain that the provisions of the Ordinance and the rules made thereunder are being complied with. A licensee is required to submit to the Minister, maps, plans and drawings on a detailed scale when the work involves dams, canals, reservoirs, or tunnels. The Pangani River Ordinance regulates the use of certain parts of the waters of the river and its tributaries. It restricts the extent of the rights which may be granted by the President or any other

---

54/ The (Pakistan) Development of Industries (Federal Control) Ordinance, 36 of 1972, amended by 37 of 1974.

55/ The Electric Power Act, Chapter 314, Laws of Kenya Section 135(1).

authority charged with the duty of controlling or regulating the river waters. The Ordinance provides that neither the President nor any such authority may allow the use or obstruction of the river waters in such a way as to reduce the flow to less than 450 ft^3/sec.

In India, the state government controls the supply of electricity; the state Electricity Board acts as the medium between the government and the users. In addition to the powers devolved on the licensees to distribute electricity, the Boards themselves may become heavily involved in water power, which is particularly relevant to the coastal regions. The Board is empowered to take measures to advance the development of water power in the state (this would include tidal power) and to organize and carry out power and hydrometric survey work necessary for such purposes. It prepares schemes for generating stations. The Central Electricity Authority recommends to the Board any scheme for developing electricity but must ensure that the power generated will not affect the natural flow of the river and will be consistent with the requirements of irrigation, navigation, and flood control. The state government alone grants concessions for development of water power and its use for electrical purposes; the state government also makes rules governing these matters, and the Board functions under its directions on questions of policy. The legal framework is similar in Pakistan, 56/ Burma, 57/ and Sri Lanka. 58/

In Mexico, the transmission, transfer or distribution of electricity are functions performed by the State; concessions to non-nationals are forbidden. 59/

All activities relating to the aplication of nuclear energy are generally under government control in most developing countries, e.g. to mine and dispose of uranium, to produce, develop, and use nuclear energy, and to manufacture, produce and buy radioactive substantances. In India and Pakistan, government is also empowered to declare 'restricted information' and 'prohibited areas'. The production and supply of electricity, the erection of buildings, and the execution of work connected with nuclear energy all fall also within the exclusive competence of government. 60/

In Peru, an exclusive right to explore, exploit and refine radioactive substances is given to the Institute of Nuclear Energy, a government agency. 61/ Both Brazil and Venezuela assigned this task to specialized agencies. Brazil created a mixed private-public company whose stock must be at least 51 per

---

56/ For example, the Act of 1910, as amended in 1940 and 1964, and the Electricity Control Ordinance, 28 of 1965.

57/ For example, the Electricity (Control of Undertakings) Act, 63 of 1947.

58/ For example, Cap. 205 Laws of Ceylon provides the framework whereby the Minister grants licences to supply electrical energy.

59/ Constitution (1917 as amended 1972) Article 27.4.

60/ India, the Atomic Energy Act, 33 of 1962 as amended, Section 3; Pakistan, Atomic Energy Commission Ordinance, 17 of 1965, as amended by Act 13 of 1968 and 16 of 1974.

61/ Decreto Ley, 21297 of 1975, Articles 1 and 2.

cent owned by the National Commission on Nuclear Energy. 62/ The corporation is also to ensure adequate treatment and elimination of nuclear wastes.

62/ Law 5740 of 1 December 1971, Article 1.

CHAPTER 13

Special Measures for Critical Sections of the Coastal Area

Some countries have selected certain critical sections of their coasts and prepared special plans for those sections. They serve the purpose of pilot projects, illustrating how planning and management techniques can be applied. The efforts made in Spain, Israel, Togo, Indonesia, Yugoslavia and Scotland are examples of this approach.

Spain recently prepared an Indicative Plan for Uses of Publicly-owned Coast from Torredembarra to Cambrils (Tarragona). 1/ The total area is 45 km, covering urban settlement, tourist centres, fishing and recreation areas. The techniques used for the preparation of the plan are worthy of brief mention. 2/

Detailed studies were first carried out regarding the various aspects of the coast, including physical aspects of the coastline, its lithological composition, current uses of the publicly-owned land, population distributions (between permanent and transitional population, and between coastal and inland population), access to the coast from public roads and promenades, and an inventory of beaches. Each of these aspects was examined in connexion with the dynamics of the coastline and an analysis of the current conditions of each particular stretch regarding the capacity of the beaches and existing installations. In evaluating the beaches' capacity and current needs, the first 10 m from the water were treated as the active area, the next 30 m as the resting area, and the remainder for setting up beach facilities.

On the basis of these studies, a master plan for future uses and needs was then prepared. The plan dealt with the various issues of forecasted uses, necessary access roads to the various sectors of the coast, promenades, shipping facilities or extension of facilities, beach protection, regeneration capability, areas for improvement. Three planning phases were used - immediate (1975), medium-term (1975-1979, coinciding with the completion of the fourth Development Plan), and long-term (to 1985). The whole coastal area

1/ The information is based on a report submitted by the Government of Spain to an interregional seminar sponsored by the United Nations Ocean Economics and Technology Office on development and management of resources of coastal areas held in Berlin (West) in June 1976.

2/ Information referred to here is also partly based on communication received from the Government of Spain, dated 14 January 1976.

under the plan was divided into zones according to the proposed uses: (i) the tourist and recreation zone, dealing with the predominant use in the area, covers not only sectors already devoted to this purpose, but also includes places currently described as having "no specific use"; (ii) the industrial zone covers basically the area south of the port of Tarragona; no industrial installations are permitted outside this area; (iii) port zones include Tarragona (commercial port), Salou (recreation port) and Cambrils (fishing and recreation port). Plans are being made to expand the port of Tarragona as far as Salou and to build two additional recreation ports along the coast; marinas for small pleasure craft are also included; (iv) zones for free use or development: these are marked in order to encourage initiatives; (v) nature reserve areas are created to keep them in their wild state; access roads are planned.

On the basis of the plan, indicative maps (scale 1:5000) were prepared by using aerial photographs. Each map contained not only basic general information (beaches, wooded areas, highways, access roads etc.), but also symbols indicating specific aspects, e.g. existing uses, natural features, lighology and stability of the beaches, and projected uses. In this manner, the interrelationships of the conditions and the various activities are visually discernible. Mistakes, if any, can be easily checked; information acquired is put to use to the fullest extent.

In view of the rapid increase in the number of tourists visiting the Spanish Mediterranean coast, and in particular, to meet the pressing need for drinking water, sanitation and the treatment of sewage resulting in certain parts of the coast from tourists, Spain recently prepared Sanitation Infrastructure Plans for tourist areas. While the plans were originally designed for the Mediterranean and Balearic coast, they have now been extended to all areas where tourists are concentrated. At present, there are 24 plans, 12 of which relate to the Mediterranean coast from the Costa Brava to the Costa del Sol, including the Balearic islands. These plans were formulated by the Ministry of Public Works in co-operation with provincial and local authorities, and approved by the Interministerial Commission for Provincial Plans.

Each plan deals with a specific region or area regarding drinking water supplies, and the evaluation and treatment of sewage. The proposed solutions vary according to the characteristics of each specific area, population factors and facilities for obtaining water resources and for drainage. The costs form the economic basis for the exploitation. In principle, they must be sufficient to cover both the financial charges arising from loans and the cost of maintaining and operating the facilities. They should also, in addition, facilitate the creation of economic conditions which will enable the service to finance future expansions on its own, so that needs may be dealt with as they arise, thus avoiding the recurrence of any lag between the time a need is identified and the time when it can be met.

In Israel, in response to the need to preserve the coast and antiquities, to control development, and to ensure public access to the sea-shore, 3/ an Outline Scheme covering an area of 300 km^2 along the coastal strip was established. 4/ The Outline Scheme is binding on all authorities and agencies

---

3/ See Planning and Building Law 5725, 1965.

4/ Outline Scheme of Israel Coastal Strip, Ministry of the Interior, Planning Department, Jerusalem, 1974, 50 pp (in English and Hebrew).

whose activities are in any way connected with physical planning in that coastal strip, to ensure that the development of the coast is consistent with the directives of the national planning bodies, and to prevent haphazard decisions based on local considerations only. While some of the techniques used are based on traditional town and county planning, the scheme is instructive in its overall approach.

The whole coast is demarcated into various areas according to their existing and proposed uses, e.g. residential areas, semi-urban residential areas, central business districts, villages and cemeteries. Different conditions are laid down for each of these categories. Special areas are assigned for national parks, nature reserves, afforestation, bathing beaches, coastal reserves, hotels, tourism, sports centres, mineral water resorts, jetties, antiquities and archaeological sites. In addition, some areas are designed for communication or for factories or workshops (industrial zone), engineering plants, power stations, salt plants, reclamation, and agriculture.

The Scheme was established on the basis of certain guiding principles: preservation of the natural coastline and safeguarding the view of the sea, free access to all parts of the shore for purposes of bathing, rest or recreation; preservation of part of the coast as national parks, nature reserves or sea-shore reserves; preservation and restoration of antiquities; allocation of areas for engineering and defence projects which have to be situated near the sea; ensuring that main roads are not built near the coast; preventing the location of industries liable to pollute the sea near the coast; grouping of hotels in selected localities along the coast, and not in every town and near every settlement; rational dispersal of land uses along the entire length of the coast, with the aim of preventing undue concentration in the central sector; prevention of residential development along the sea-shore, even within municipal areas. 5/ Since the Scheme cannot allocate enough areas along the Mediterranean and the Elat coasts to satisfy all requirements, a division has to be made; thus, for example, 50 per cent of the Mediterranean coast was allocated for bathing beaches, 15 per cent for national parks and nature reserves, 4 per cent for engineering projects, 5 per cent for industry and transport, and 24 per cent for other uses.

A draft law for the management and development of the Togolese coastal zone is being considered in connexion with the five-year plan. This proposal deals with the development aspects of the coastal zone in the light of Togo's coastal needs and problems, as for example roads, coastal highways, erosion (particularly near Lomé), tourism, the creation of urban zones for petroleum industries and the preservation of forest and agriculture. To ensure recreation and public access, protection zones will be established along the coastal line within a strip of 100 m width along with the major river mouths, lakes, ponds, and lagoons. 6/

5/ See E. Efrat. Physical Planning Aspects of Israel's Coastal Strips, Ministry of the Interior, Planning Department, Jerusalem, 1973.

6/ An economic background information on the coast of Togo may be found in Y. Aziaha's contribution made to the Interregional Seminar on Development and Management of Resources of Coastal Areas, Berlin (West). In K. H. Szekielda and B. Breuer (Eds.), Development and Management of Resources of Coastal Area, 1977 published jointly by German Foundation for International Development and the United Nations, 546 pp.

The province of North Sumatra (Indonesia) has adopted a Coastal Area Development Plan for 1975-1978. 7/ It is basically an economic development plan for the coastal villages on the east coast (Straits of Malacca) and on the west coast (Indonesian Ocean).

The basic policy is to develop the maximum potential of the region. The principal development activities are sea and inland fisheries, aquaculture, farming (tidal irrigation of rice-fields), coconut plantation, livestock breeding and home industry. Development funding is being made available and is to be distributed to the region according to the density of population. There is a guiding team which is responsible for the implementation of the plan.

Detailed implementation is set out in the plan for each development activity. Thus for example, regarding smallholder farming, the means include opening up tidal rice-fields, providing guidance in fieldwork, providing training and technical cultivation courses, developing tools for processing land and for crop protection, providing loans to supply fertilizers, and building rice mills etc. The plan also identified areas where further research is needed.

In Yugoslavia, the pressure of tourists on the Adriatic coast has led to the destruction of natural attractions and special features of the countryside. 8/ The Yugoslav Government, assisted by the United Nations, prepared a long-range development plan for tourism in the South Adriatic region. Existing urban legislation has served as a basis for the preparation of new proposals. The concept of planning has been introduced as a continuous process to provide flexibility in the plan, identifying hierarchies and relationships between plans at different levels, e.g. regional studies, master plans and urban studies for new tourist settlements. In accordance with the plan, new tourist settlements are to be situated away from the shore and absorbed into green belts, thus providing within the area of development green areas for leisure and recreation.

The regulations suggested in the plan for the physical development of the South Adriatic coast cover not only the study and co-ordination of all legal matters necessary for implementation of the proposals, but also legal measures to ensure the protection of the human environment as an integral part of the proposals. Since three republics are involved in the plan, a special Inter-republican Co-ordinating Committee was created to assume a supervisory function.

In Scotland, a study of the coast by the Scottish Development Department with the help of other agencies, resulted in the preparation in 1974 of coastal planning guidelines for North Sea oil and gas development. These

---

7/ Coastal Area Development in the Province of North Sumatra, the Three-Year Programme, 1975-1978, Badan Perencenaaan Pembangunan, Propinsi Sumatra Utora, 1975. Development Planning Agency, North Sumatra Province, 1975, 197 pp. The actual implementation of the plan has been delayed.

8/ See Car, Cicvarié, Radic, Sekulié, "De quelques effets économiques du tourisme étranger en Yugoslavie", in La planification et le développement de l'industrie touristique dans la région de la CEE, CEE, New York Acte du Colloque 13-18 Octobre 1975. ECE/SEM.1/4 pp. 152-167.

guidelines are intended for the use of planning authorities in dealing with demands on the coastal areas for oil and gas development. 9/

Their basic principle is that exploitation can only take place in those areas designated for that purpose. This is to avoid a scatter of industrial development, to permit full use of existing labour, housing and public services, and to ensure the possibility of diversification to cushion any subsequent decline. These guidelines appear to be established on the basis of economic, physical and environmental considerations. The guidelines set out various criteria for establishing preferred zones for development or conservation. The preferred development zones are those areas that are able to be expanded without incurring the risk of economic or social decline; that have some flat land on the coast and in the hinterland able to absorb major development; that possess suitable ports and harbours with some potential for developing the dockside land; or in which environmental recovery and rejuvenation are feasible. The preferred conservation zones are those places vulnerable to development because of their scientific, ecological or scenic features; sections of the coastline where an existing or proposed use (bathing) would be incompatible with major oil and gas development; areas of coast containing small scale communities where expansion might cause serious economic and social problems. Commercial development is also excluded from the tourist and recreation areas, and areas of the coast with towns and villages whose historic character should be protected. These criteria can be usefully considered for countries needing guidelines for development in the coastal area.

9/ North Sea Oil and Gas: Coastal Planning Guidelines, Scottish Development Department, P/HD/29/12/7/1, 15 pp.

CHAPTER 14

Establishment of National Guidelines for Management Programmes

It is generally recognized that the main difficulty lies in providing a management programme at the national level for the purpose of governing the use of each individual track of coastal land or waters. This may be even further complicated by the administrative set-up, particularly if the responsibilities are divided between national government, state government and numerous local units of government, such as cities and counties. These were some of the main reasons why the United States chose to adopt national guidelines for a management programme, leaving the power of implementation to local government to allow diversity. Its 1972 Coastal Zone Management Act 1/ thus deals with the process of developing a programme through the establishment of various elements which a coastal state must examine in structuring its management programme. These elements include the determination of permissible uses and priorities among uses, the types of manpower and programmes required, the identification of areas requiring special attention, and co-ordination among the agencies and units of government at all levels with responsibility for and concerned with the coastal area. The Act focusses on the process and leaves the substance of the management programmes for the states to define, provided that the state programmes are committed to the management of the coastal area on a co-ordinated and comprehensive basis.

The system of government in the United States is a three-tier one which divides responsibilities between national government, state government and thousands of local units of government, such as cities and counties. A deliberate effort is therefore made in coastal management to co-ordinate the relationship among the three levels of government. The federal role is carefully restricted to encouraging the states through providing financial

---

1/ For an analytical treatment of the process of programme development on the basis of the Coastal Zone Management Act of 1972, see Armstrong, Bissell, Davenport, Goodman, Hershman and Sorensen, Coastal Zone Management: The Process of Programme Development, Coastal Zone Management Institute, 1974; Coastal Zone Management in the U.S., UN document E/CONF.70/TP/190, January 1977, paper submitted bo the UN Water Conference held in Argentina, March 1977; R. W. Knecht, "Coastal Zone Management: A Federal Perspective", Coastal Zone Management Journal, 123, (1974). For a survey of coastal state legislation in the U.S., see J. M. Robbins and M.J. Hershman, "Boundaries of the Coastal Zone: A Survey of State Law", ibid., p. 305.

assistance and guidelines on what their coastal programmes must contain in order to qualify for continued federal funding. However, the main responsibility for coastal area management is given to the state. State governments are required to work together with local units in the coastal areas. At the same time, state and local officials are also required to consult closely with those federal agencies operating on the coast in the preparation of their programmes.

State management programmes are to be based on a comprehensive strategy to guide future decisions on what is to take place in a coastal area. While programmes naturally differ from state to state, they generally aim to strike a balance between development needs and preservation requirements. One of the federal requirements is that states designate coastal areas of "particular concern" on the basis of their environmental sensitivity. Special regulation might be called for in such areas, and also a permit for any development activity which is to be considered at state level instead of at local level only - the latter being the normal practice. State programmes are also required to identify which areas of the coast are considered suitable for industrial or large commercial development. By declaring in advance which areas are subject to special regulations and which areas are suitable for development, both preservation and development interests are served. Disordered development is thus prevented.

The preparation of a coastal management programme may take many years, for states need to collect basic data about their coastal areas, to identify the necessary agencies under state law to carry out a programme, and to settle such main issues as the boundary of the states' coastal zones. The provision of financial assistance for developing such a programme is an important objective of the Act. Federal government usually finances up to 80 per cent of the cost of the preparation of a programme. Once a state has successfully completed the preparation of a programme, setting out its goals and the means it has selected for achieving them, the programme will then be submitted for approval. Once approved, the programme qualifies for full funding. Sometimes an interim implementation phase may be necessary for states which need to pass legislation to give the coastal authority sufficient legal basis to meet the requirements.

One of the difficulties in connexion with coastal programmes in the United States relates to private property development, which is a source of potential conflict in many states. While most property owners recognize the right of local government to regulate property use, some have resisted the idea that a measure of this control be exercised at the state level. Efforts are being made to maintain a balance which would respect the rights of private owners but at the same time recognize the interest of government in regulating private property use so that environmental or social damage is not caused.

Although the Coastal Zone Management Act itself has an overtone of environmental concern, the recent economic situation in the United States has led to an emphasis on economic development and expansion as well, thus, making the implementation more realistic than it would have been otherwise. At present, all 30 coastal states and three of the four territories eligible have introduced substantive management programmes and/or legislation. These programmes and legislation are shaped in the light of their perceived needs and the geographical characteristics of the different states. 2/ On the

2/ Status of State Coastal Zone Management Efforts, U.S. Department of Commerce, 1975.

Atlantic side, much attention has been focussed on protecting wetlands because of their tremendous value to coastal fisheries and other estuarine life, and the danger to such estuarine areas from dredging and filling projects and other coastal developments. 3/ The Gulf coast states have concentrated on methods for the allocation of exploitation rights, and have only begun to implement protective measures as the effects of overdevelopment have begun to threaten their coastal resources. 4/ In the Great Lakes area, shoreland management programmes have been the major concern primarily in response to pressure on their inland lakes. 5/ On the Pacific coast, 6/ the emphasis has been placed largely on regulating the use of beach areas. Thus, the legislation adopted by the various states covers a wide spectrum of concerns. 7/ As to the administrative machinery, the solutions adopted vary greatly, ranging from a total administrative reorganization creating a single state agency having comprehensive jurisdiction, to interagency groups. Again, the administrative machinery selected represents what was considered most appropriate and necessary for a given case. An examination of the existing programmes shows that five distinctive approaches may be identified:

(i) Comprehensive planning: more than 15 coastal states have completed or are undertaking major studies of their coastal resources and on methods of their effective utilization. The results of the studies have been or will be

---

3/ See for example, Report of the Conference on Marine Resources of the Coastal Plains States, December 1977, Williamsburg, Virginia, published by Coastal Plains Center for Marine Development Services.

4/ See for example, Comparative Aspects of Coastal Zone Management: Background Information on the Law of Texas and Other States in View of the Coastal Zone Management Act of 1972, Bates College of Law, University of Houston, 1973; E.G. Fruh, Establishment of Operational Guidelines for Texas Coastal Zone Management, University of Texas, 1974; Alabama Coastal Area Programme, Alabama Development Office, 1974.

5/ See Annual Report for 1976, Great Lake Basin Commission, 1977; J.M. Armstrong, Coastal Zone Management in the Great Lakes, paper presented to the 14th Conference of the Conference for Great Lakes Research, Toronto, 1971.

6/ See for example, Statewide Planning Goals and Guidelines, Adopted by the Land Conservation and Development Commission, 1974, Oregon Land Conservation and Development Commission, 1975; Recommendations for Development Activities in Florida's Coastal Zone, State of Florida, Department of Natural Resources Coastal Coordination Council, 1973; Suggested State Objectives, Policy and Criteria for Coastal Management in Florida, Florida Bureau of Coastal Zone Planning, 1977.

7/ The Seventh Annual Report of the Council on Environmental Quality September 1976, pp. 67-70. For a detailed analysis of these programmes and legislation, see Bradley and Armstrong, A Description and Analysis of Coastal Zone and Shoreland Management Programmes in the United States, University of Michigan, Technical Report No. 20, March 1972. About 14 states wished to protect wildlife and fishery, 12 states to preserve or protect coastal eco-systems, 11 states to enhance aesthetics, 9 states to protect water resources and to conserve soil resources; 12 states have laws to regulate coastal growth and development; see Zwicky and Clark, "Environmental Project Motivation in Coastal Zone and Land Use Legislation", Coastal Zone Management Journal, No. 1, pp. 103-108, (1973).

used in the sequential development of a comprehensive management plan and an administrative mechanism to implement it. The planning programmes vary in their basic premises and differ in scope and sophistication.

(ii) Functional approach: Washington and Rhode Island established a functional management mechanism, which is equipped to deal with pressing developmental problems revealed by a quick inventory or survey. The activities are guided by a general statement of management philosophy and a loose and flexible set of long-range objectives. The programmes are designed primarily to give direct and immediate administrative action, and to avoid an extended pre-planning period.

(iii) Temporary moratorium: some states under strong developmental pressure, e.g. California, Delaware, Oregon, and Texas, have introduced legislation which declared a general or selective moratorium on development within a defined coastal or offshore strip during a planning period of specified length.This approach represents an attempt to control short-term development and thereby establish a more favourable climate for long-range planning.

(iv) Land zoning: some of the Great Lakes states, e.g. Wisconsin and Michigan, have chosen to encourage their county governments under the loose supervision of the state, to regulate coastal development through traditional single-use land zoning. The counties have been given a specified period to establish zones for certain permissible uses in a narrow coastal strip.

(v) Inclusion of coastal planning: some other states, e.g. New York and Maryland, chose to expand the responsibilities of the existing administrative organs to include coastal planning and, to a lesser degree, management functions. They hoped thereby to avoid the normal frictions created by the establishment of a new bureaucracy within the existing administrative structure. The State of Hawaii took a similar approach. It had already instituted a statewide land use control programme and, after further investigation, it was decided to introduce in addition a coastal zone programme which covered such problems as shoreline setback, beach access, marine management and protection of natural areas.

The different approaches adopted in the United States are illustrations of some of the basic methods which could be used to suit different geographical locations, administrative set-ups and perceived needs.

Marine uses and resources are playing an increasingly important role in national economies. In the last few years, countries have established their sovereignty or jurisdiction over resources and uses of the ocean generally to 200 miles. Some are taking steps to introduce the "marine dimension" into their national economic planning and development. The establishment of a national management and development programme requires an adequate scientific data base, trained personnel, an administrative and institutional framework, and access to appropriate technology. The fulfillment of these requirements for most countries calls for collaborative effort at the national and international levels. It would appear that interested countries, and institutions having special responsibility for marine resources and uses, would greatly benefit from the exchange of information and experience in carrying out this new endeavour.

CHAPTER 15

Measures Taken by the United Nations

1. Introduction

Within the United Nations, the process of involvement in marine affairs was initiated in 1966 when the Economic and Social Council considered that the mineral and food resources of the sea beyond the continental shelf, excluding fish, constituted reserves of raw materials which were as yet not being fully utilized, and that the rational use of these resources to ensure optimum yield and minimum waste was of vital importance to all countries. In the same year, the General Assembly also noted that the effective exploitation and development of ocean's living and mineral resources could raise the economic level of peoples throughout the world, and in particular that of the developing countries.

In 1968, the Economic and Social Council recognized that further progress in the development and utilization of marine resources would depend greatly on more specific knowledge being gained on the various characteristics of marine environmental conditions, and observed that one of the main obstacles to the development of marine science and technology was the shortage of experts and qualified personnel, particularly in developing countries. The General Assembly also considered it important to promote international co-operation for the exploration and exploitation of ocean resources and that it was essential to provide within the United Nations system a focal point, both for the elaboration of desirable measures of international co-operation, taking into account alternative actual and potential uses of the ocean, and for the co-ordination of the activities of international organizations in that regard.

In 1972, the General Assembly underlined to coastal states the importance, for purposes of economic development and social progress, of the ocean resources adjacent to their coasts. In 1973. the Economic and Social Council called attention to the fact that the coastal area in many countries, particularly in the developing countries, represented one of their most valuable possessions, and that its proper management and development constituted an important factor in their national development planning. The Council also pointed to the need for an integrated approach in the consideration of the problems involved in the various uses of the sea, to achieve proper management of the ocean's resources and uses, and for an integrated information base to support such an approach.

Two years later, reaffirming the importance of an integrated approach to coastal area development, the Economic and Social Council in Resolution 1970 (LIX) endorsed a work programme in this field and underscored the need for measures to promote the better and wider application of marine technology suitable for developing countries.

Most recently, in 1977, the Council by Resolution 2099 (LXIII) provided further impetus to the collaboration between the United Nations, the Intergovernmental Oceanographic Commission and the Food and Agriculture Organization in the development of the Aqutic Sciences and Fisheries Information System (ASFIS) and to the elaboration of activities in the fields of coastal area development and marine and coastal technology.

2. Preparation of Proposals and Methodology

Activities in the field of coastal area management and development were launched in 1973-1974 with the preparation of a comprehensive conceptual study and the convening in November 1974 of a meeting by a Group of Experts on Coastal Area Development in order to assist in preparing practial proposals for action at the regional and subregional levels. 1/ At this initial stage, several regions were identified as being suitable for efforts to foster a co-operative approach to the opportunities and problems presented by the exploitation of marine and coastal resources and the use of the coastal area. These regions included the Persian Gulf, the Gulf of Guinea, the Caribbean Sea and Southeast Asia.

During the same period, the United Nations Environment Programme (UNEP)undertook a major project for the countries of the Mediterranean Sea, and this "Regional Seas Programme" approach has been extended to other regions including those mentioned above. As a consequence, these two complementary approaches have been linked as have other United Nations activities to form a comprehensive systemwide programme under UNEP co-ordination.

An Interregional Seminar on Development and Management of Resources of Coastal areas was convened jointly with the German Foundation for International Development in Berlin (West), Hamburg, Kiel and Cuxhaven from 31 May to 13 June 1976. 2/

Representatives from some 30 developing countries, assisted by resource specialists from the United Nations system as well as from industry and educational institutions, examined the problems and opportunities associated with coastal resource development and management with a view to developing concrete action proposals that would receive expression in follow-up activities at the national, regional or subregional levels.

1/ For the study, a draft of which was submitted to the IOC Secretariat and other United Nations organizations for critical review, see E/5648, 8 May 1975, Coastal Area Management and Development: Report of the Secretary-General.

2/ The proceedings were published in 1977: Development and Management of Resources of Coastal Areas published jointly by the United Nations and the German Foundation for International Development, 1977, Dok.855 A/a III-S 14/76, 546 pp.

The format of the Seminar was structured in such a way as to encourage action-oriented results within an integrated planning framework. Thus, the participants were first exposed to lectures and discussions ranging from the need for integrated planning to artificial beaches and coastal mapping from satellite altitudes. They were then organized into concurrent workshops that focussed on specific priority themes and problems, identified on the basis of presentations of national experience.

The recommendations made at the workshop have been one source of impetus to more specific regional activities that are now being implemented under the ocean economics and technology programme.

These early activities were aimed at creating an awareness of and familiarity with concepts and problems and activities associated with an integrated approach to coastal area development, and establishing a co-oprative framework for a systemwide effort. They, in turn, have provided the basis for the elaboration of more action-oriented measures in the form of (1) studies on the components of an integrated approach to coastal area management and on particular problems (erosion) or physical contexts (island countries), and (2) practical assistance in the areas of training and substantive support for technical co-operation (workshop, field projects).

Preliminary work is also presently underway for the preparation of a study on marine resources, energy, and alternative development strategies for island countries. Starting with the premise that small islands may lack the minimum critical mass required to pursue the export-oriented, industrial development strategies encouraged under traditional development models, the study will explore the possibilities of more 'inward'-looking strategies in relation to the marine resource-related sectors. With an initial focus on the Caribbean, this work is related to other activities being carried out in that region.

3. Four Coastal Regions

As already mentioned, regional and subregional activities are also being implemented in the four coastal regions initially selected for special attention.

(1) The Caribbean

Caribbean Development and Co-operation Committee (CDCC). At its first session in 1975, the CDCC of the Economic Commission for Latin America adopted a work programme that included a section on coastal area development. In response to this expression of interest by governments in the Caribbean subregion, which took full account of the recommendations by the Group of Experts on Coastal Area Development concerning the region, the Ocean Economics and Technology Office prepared a draft background study entitled "Development and Environment in the Caribbean: coastal and marine aspects". 3/ The study was intended to provide a comprehensive review of the current situation in the Caribbean with respect to marine resource development and potential, uses of coastal and marine space and the state of the marine environment. A considerably abridged version of the lengthy study was subsequently prepared for use by an Interagency Meeting on Co-ordination for Implementation of the Work Programme of CDCC, convened in June 1976.

3/ The study was mimeographed.

The Ocean Economics and Technology Office also prepared a further study entitled "Coastal Area Development and Environment in the Caribbean: coastal and marine aspects," 4/ and submitted it to the CDCC at its fourth session in Surinam in March 1979. The study provided a review of the importance of coastal and marine-related activities by subregions. It also contains specific programmes which might be undertaken at the regional, sub-regional and national levels, to promote the development of coastal areas and marine-related activities in the Caribbean. Among the sectors and problems covered were training, research and supporting services; protection from natural hazards; marine pollution; marine transport; fisheries; and tourism. The CDCC agreed to the proposals made. 5/

IOCARIBE. At its first session held in Caracas in July 1976, the Intergovernmental Oceanographic Commission's Association for the Caribbean and adjacent regions (IOCARIBE) took note of the "development programme of the . . .Ocean Economics and Technology Office (of the Department of Economic and Social Affairs) in the Caribbean, particularly as it related to training and education, coastal area development and management and the IOCARIBE project on environmental geology and evolution of the continental and insular margins of the Caribbean and Gulf of Mexico area". IOCARIBE further considered it "essential that it work closely with . . . the Ocean Economics and Technology Office and UNEP" and requested "the IOCARIBE Secretariat to establish open channels of commuinication particularly with these two bodies. . ." In line with this recommendation, the Ocean Economics and Technology Office contributed to the IOCARIBE Environmental Geology Workshop convened in January 1978 in Port-of-Spain, Trinidad and Tobago, and has been co-operating closely with the IOCARIBE Secretariat in organizing a regional training workshop on coastal area management and development. 6/

Action Plan for the Caribbean. At the request of UNEP and within the framework of its Regional Seas Programme, the Ocean Economics and Technology Office in 1979 prepared a study on marine and coastal area development in the wider Caribbean. This study examines the patterns of economic activity in the Caribbean and its effects on the marine environment. The implications of future growth are also examined, and recommendations are made for action at national and international levels. 7/ The study has served as a major input into an experts' meeting convened in the region by UNEP in early 1980. Following the pattern established for other regions, the experts' meeting is one of the several steps leading to a plenipotentiary conference in the region which will adopt an action plan.

(2) The Gulf Region

The Gulf Region was designated as one of four to receive priority examination. A plan for a multi-disciplinary expert mission was designed at the meeting of the Group of Experts on Coastal Area Development. The IOC and other specialized agencies of the United Nations system were requested to participate to ensure an integrated nature of the project.

4/ Document E/CEPAL/CDCC/52/Add.1, 19 March 1979.

5/ Document E/CEPAL/CDCC/54/Rev.1, Vol. I, para. 183.

6/ See below.

7/ The study will be issued as a joint UN/UNEP/ECLA publication.

In early 1976, the United Nations Environment Programme agreed to undertake a major role in project financing. Following briefings at headquarters and at Geneva in March 1976, the mission spent two months in the region, visiting each of the eight participating countries (Iraq, Iran, Kuwait, Saudi Arabia, Bahrain, United Arab Emirates, Qatar and Oman).

The working papers prepared by the team were compiled and edited in June 1976. A summary report of these, incorporating agency comments and suggestions, was prepared in August 1976, and together with a draft action plan, was submitted to an experts' meeting on coastal area development and protection of the marine environment jointly sponsored by the Department of Economic and Social Affairs of the United Nations Secretariat and UNEP, and held in Kuwait from 6 to 9 December 1976.

This was fllowed by a further experts' meeting in Nairobi in 1977, which led to the Kuwait Regional Conference of Plenipotentiaries on the Protection and Development of the Marine Environment and the Coastal Areas held in April 1978. The Conference adopted an Action Plan for the region, a Convention for "co-operation on the protection of the marine environment from pollution", and a Protocol concerning "regional co-operation in combating pollution by oil and other harmful substances in cases of emergency". In view of the priorities and needs of the region, the following co-operative programme relevant to the management of regional environmental problems stemming from national development activities are being undertaken:

(a) Formulation of regional contingency plans for accidents involving oil exploration, exploitation and transport, and the strengthening of the meteorological services contributing to the development of contingency plans and to their execution in co-ordination with existing or future marine regional meteorological programmes;

(b) Assistance in the development of national capabilities in engineering knowledge needed for regional environmental protection;

(c) Strengthening the national public health services and their co-ordination whenever transboundary interests require it;

(d) Rational exploitation and management of marine living resources, including aquaculture, on a sustainable basis, and the establishment of protected aquatic and terrestrial areas, such as marine parks, wetlands and others;

(e) Co-ordination of marine and land transport activities and the creation of a regional transport co-ordinated programme with special emphasis on port-generated pollution;

(f) Development of principles and guidelines for coastal area development and management through workshops;

(g) Co-ordination of national water management policies including community water supply and water quality control, wherever they may have an impact on the marine environment of the region; and

(h) Upkeep of records of oil pollution incidents in the region with relevant information on the impact of such pollution on the marine environment.

According to the Convention, the state parties undertake the general obligations to prevent, abate and combat pollution of the marine environment, resulting from land-based and sea-based sources; 8/ co-operation in dealing with pollution emergencies; liability and compensation; and the establishment of the Regional Organization for the Protection of the Marine Environment to be located in Kuwait. This organization consists of a council which will convene annually, a secretariat to service the organization and a judicial commission for the settlement of disputes.

Under the Protocol concerning "regional co-operation in combating pollution by oil and other harmful substances in case of emergency", the participating countries undertook to co-operate in taking necessary and effective measure to protect the coastline from the threat and effects of pollution from oil or other harmful substances resulting from marine emergencies in the sea area which includes ports, harbours, estuaries, bays and lagoons. The Protocol provides for the establishment of the Marine Emergency Mutual Aid Centre which will have primarily a co-ordinating role in the exchange of information, training programmes and monitoring and which will also assist in the preparation of laws and regulations and the design of marine emergency contingency plans. Because of its expertise and considerable experience in these matters, IMCO will assume responsibility for organizing the Centre. The possibility of the Centre initiating operations to actually combat pollution at the regional level may be considered at a later stage.

Each contracting State agrees to require all appropriate officials to report the existence of any marine emergency to the national authorities who will, in turn, inform the Emergency Mutual Aid Centre, all other contracting States and the flag State of any foreign ship involved in the marine emergency concerned. The parties to the Protocol further agreed to assist one another in responding to emergencies, including through personnel, material and equipment, surveillance and monitoring or the mobilization of other pollution combating resources.

The United Nations Environment Programme has assumed the responsibility for providing the interim secretariat and for the administration of the Regional Trust Fund which was established to cover the costs of implementing the Action Plan. This Trust Fund is financed for the initial two and a half year period by proportional contributions from the governments totalling U.S.$5.82 m with U.S. $500,000 contributed by UNEP towards the cost of the interim secretariat.

In the follow-up to the conference, the Ocean Economics and Technology Office has prepared operational plans for the projects under the Action Plan for which it has been assigned a lead agency role. These projects, all of which require inputs from other United Nations organizations, pertain to the assessment of geologic processes contributing to, or modifying the fate of pollutants in the region, and their impact on human health, marine ecosystems and human activities, as well as of the effects of coastal engineering and mining; the survey and identification of major development activities and trends of their future development; the co-ordination of marine and land transport activities and the creation of a regional planning and co-ordination programme with special emphasis on port-generated pollution; and the development of principles and guidelines for coastal area development and management through workshops.

8/ See Final Act of the Kuwait Regional Conference of Plenipotentiaries on the Protection and Development of the Marine Environment and the Coastal Area, UNEP Publication, 1978.

(3) Southeast Asia

As in the case of activities in the two regions described above, the 1974 Group of Experts Meeting on Coastal Area Development provided a springboard for a parallel approach in Southeast Asia. In December 1979, a workshop on coastal area development and management was organized in Manila. Contributions to the workshop are being provided to ESCAP by the OETO with support from the German Foundation for International Development, the East-West Center in Hawaii, UNEP, and the Committee for Co-ordination of Joint Prospecting for Mineral Resources in Asian Offshore Areas (CCOP). The participating countries included the members of CCOP as well as those of the Committee for Co-ordination of Joint Prospecting for Mineral Resources in South Pacific Offshore Areas (CCOP/SOPAC). United Nations experts and consultants in certain specialities assisted the national participants in examining data collection and analysis techniques, planning methodologies, research programme design, and administrative regulation procedures.

(4) West Africa

As a follow-up to the meeting of the Group of Experts on Coastal Area Development, the Ocean Economics and Technology Office prepared a paper entitled "Preliminary study on the major uses of the coastal area development scheme", 9/ which was transmitted to UNDP and the Economic Commission for Africa (ECA) for comment at the end of 1975. This study was among the sources of information considered by a UNEP Exploratory Mission on Marine Pollution Problems of the West African Coastal States of the Gulf of Guinea in Spring 1976.

An international workshop on "Marine Pollution in the Gulf of Guinea and Adjacent Areas" was held in Abidjan, Ivory Coast in May 1978. This workshop was one of several preparatory activities undertaken to provide the basis for an action plan for the region.

In 1979, at the request of UNEP, the Ocean Economics and Technology Branch prepared a background study on socio-economic activities in the region relevant to the marine environment, which was used in the preparation of the draft action plan.

A regional seminar/workshop on coastal area development in West Africa has been scheduled for 1981. The workshop is expected to contribute to meeting extensive training needs that have been identified in earlier preparatory work as a priority for the region. It will not only enable decision-makers to appreciate more fully the integrated and comprehensive planning principles to be embodied in the action plan for the region, but should also highlight problems of regional concern and those commonly faced by the West African Coastal States. In the short-term, the workshop will reinforce UNEP's peparatory effort aimed at the adoption of a regional action plan and, in the long-term, facilitate the implementation of the action plan at the national level.

Recognizing the effects of erosion on coastal area development, the Governments of Benin and Togo requested the United Nations to organize a workshop on "Causes of and Possible Solutions to the Coastal Erosion Problems in Togo and Benin". The workshop was held in January - February 1979, and evaluated existing cases of erosion in Benin and Togo, identified the causes

9/ The study was mimeographed.

and possible solutions, and outlined future actions which the Governments might wish to take. National representatives of Benin and Togo together with international experts with divergent backgrounds and experience took part in the workshop. Certain plans have already been made and implemented at the national level. 10/

4. Training and Information Activities

In response to concern expressed by the Economic and Social Council training and information activities in marine and coastal affairs have been expanded over the past several years.

A first step in this direction was the publication late in 1976 of a Registry of Courses and Training Programmes in Marine Affairs, 11/ which compiled information on the number, diversity and content of courses and programmes offered by institutions throughout the world in marine-related fields. It is hoped that the Register will help the developing countries, in particular, in their efforts to develop an infrastructure of trained manpower in the marine capabilities.

As a member of the Intersecretariat Committee for Scientific Programmes Relating to Oceanography (ICSPRO), the United Nations is closely involved with the Training, Education and Mutual Assistance (TEMA) component of the Intergovernmental Oceanographic Commission (IOC) and relies on it as the main mechanism for organizing training activities in the field of coastal area management and development. At the second session of the IOC Working Committee for TEMA in 1977 12/, a resolution on "Coastal Zone Management and Marine Pollution", was adopted recommending inter alia, that "the IOC in conjunction with the ICSPRO agencies take the necessary steps to implement training courses on Coastal Area Management and Development . . . " 13/ At its tenth session, the IOC Assembly provided further impetus to this initiative. 14/

IOCARIBE, assisted by the Ocean Economics and Technology Office, in co-operation with the Government of Mexico, and with the support of the IOC Secretariat and the Division of Marine Science of UNESCO and UNEP and the United Nations University organized a training workshop on "Coastal Area Management in the Caribbean Region" in September-October 1979. The purpose of the workshop was to familiarize selected government officials having high level policy-making and planning responsibilities, with the concept of coastal area development within the programme of national development planning and to introduce them to the prerequisites of a management approach to coastal area development.

10/ The findings of the Workshop are being published both in French and English in European Applied Research Reports: A Journal of European Science and Technology, Harwood Academic Publishers in Spring, 1980.

11/ United Nations publication, Sales No. E.77.II.A.2.

12/ The UN/OETO substantive contribution is contained in the Report of the ICSPRO Agencies to the Working Committee, document IOC/TEMA-II/8, 13 June 1977.

13/ Summary Report, document IOC/TEMA-II/3, 18 August 1977.

14/ See Resolution X-18, "Financing 1977 TEMA activities" and X-19 "The Working Committee for TEMA".

5. Marine and Coastal Technology

The activities in marine and coastal technology further support the programmes in coastal area management. The objective of activities in this field is to assist developing countries in identifying and evaluating specific marine and coastal technologies most appropriate to their needs, and in strengthening their capabilities, individually and collectively, for developing or acquiring, adapting and applying such technologies.

The problems to be addressed include the neglect or inadequate treatment of certain types of coast-related problems, such as erosion, pollution and the destruction of natural resource endowment (for example, mangrove swamps); the failure to exploit or, through the application of inappropriate technologies, the inefficient exploitation of coastal area resources (in terms of cost and productivity, external diseconomies and failure to take full advantage of local resources - labour and materials). Within the wide range of secondary problems that give rise to the above situation, as for example shortage of capital and expertise, and obstacles to the transfer of technology, specific attention is directed to the following:

(i) Insufficient appreciation of the need for, and of the range of technological alternatives available for, meeting identified problems with specific and appropriate marine technologies. This situation is in large measure attributable to a lack or shortage of information and knowledge of the types and characteristics of marine and coastal technologies available on a global basis. In this connexion, there is insufficient knowledge concerning alternative suppliers of given technologies and the economic and technical criteria for making a selection among suppliers;

(ii) Too great a reliance among developing countries in meeting their identified technology needs on suppliers from developed countries and an insufficient degree of technical co-operation among developing countries for sharing resources, experience and expertise.

In 1978, the Aquatic Science and Fisheries Abstracts (ASFA), which is a joint project of the United Nations, FAO and the IOC, was broadened to include ocean technology, policy, and non-living resources (ASFA-II), in addition to biological sciences and living resources (ASFA-I). As an input centre for ASFA-II, the United Nations Ocean Economics and Technology Office abstracts and indexes relevant articles from some 20 periodicals as its monthly contribution to the publication, which is disseminated to some 800 institutions catering to primary users around the world.

A guide to sources of information and assistance in marine and coastal technology, (publications, fellowships, training, finance, equipment) within the United Nations system is under preparation. By providing a central source of information that is presently widely dispersed, this guide is intended to facilitate access of developing countries to the resources, expertise, information and services available within the United Nations system. During the 1980-1981 period, a second guide will be published to include sources of assistance and information outside the United Nations system, primarily those of industry and government.

A series of technology reviews is planned, each review dealing with a specific marine and coastal activity or problem. Each will provide a state-of-the-art analysis of the techniques/technologies available to deal with the given problem or activity, and compare costs of alternatives with a view to pinpointing the technical feasibility of the substitution of low-cost factors for high-cost ones. The objective in each case will be to identify the technologies that may be considered appropriate for developing countries on the basis of their factor mixes. Information will be provided on alternative suppliers and on criteria for selection. Two reviews will be published in 1981 on the subjects of "coastal erosion control" and "coastal resources inventory" and others are planned.

In order to encourage co-operation among developing countries in the marine and coastal technology field, a survey is planned of the marine capabilities of developing countries - that is, present technology applications; activities and programmes of governments, universities and commercial organizations; consultant/expert resources; and physical facilities.

6. UNEP Activities in the Mediterranean

Under the sponsorship of the United Nations Environment Programme (UNEP), the countries bordering on the Mediterranean have drawn up three regional instruments designed to combat marine pollution. 15/ For each particular source of marine pollution, a number of specific actions are contemplated, including the preparation of additional protocols to be associated with the regional instruments. The question of land-based sources of pollution was considered by an intergovernmental group in 1977 and is now ready for government approval. It is expected that eventually the provisions in these international legal instruments and the actions contemplated thereunder will be transformed into national law.

A series of studies by national research institutions in the Mediterranean region have been commissioned to specify likely development to the year 2000 in a variety of sectors, including urbanization, industrialization, agricultural development, transportation (maritime and land), offshore exploration and exploitation, and energy production and use. It is hoped that these studies, properly conducted and supported by governments, will aid national planners and decision-makers by improving the quality of information on which their actions are based. 16/ In addition, as part of the Action Plan for the Protection and Development of the Mediterranean Sea Area, adopted in February 1975, programmes on Mediterranean marine parks and wetlands have been

---

15/ These are the Convention for the Protection of the Mediterranean Sea against Pollution; the Protocol for the Prevention of Pollution of the Mediterranean Sea by Dumping from Ships and Aircraft, and the Protocol Concerning Co-operation in Combating Pollution of the Mediterranean Sea by Oil and other Harmful Substances in Cases of Emergency. See Activities of the United Nations Environment Programme for the Protection and Development of the Mediterranean Region, UNEP publication, October 1976, pp. 82-86. For a summary of the techniques used in preparing the instruments, see Drafting of Regional Legal Instruments for Marine Environment Protection: The Case of the Mediterranean, a document submitted by FAO to the UNEP Task Force on Legal Instruments for Regional Seas held in Nairobi, November 1976.

16/ M. Tolba, "La Méditerranée, une éco-région pour demain", 2000, La Documentation Française, Paris, pp. 2-7, 1976; R. Bourone et M. Tangi, "Un plan bleu pour des actions en Méditerranée", ibid. pp. 40-42.

carried out. An intergovernmental body is working on guidelines and technical principles for the establishment and management of protected areas. 17/

In response to a project concerning the Mediterranean coastal zones launched by the Organization for Economic Co-operation and Development, the Institute of Studies for Economic Planning in Italy prepared in 1974 a comprehensive study on the coastal environment in that country. 18/ The study analysed the problems in the coastal area (urbanization, industrial location, tourist development, fishing, mineral resources), coastal environment and the applicable laws. Further to that study, some long-term policies designed to integrate land use and marine resources in different areas were also proposed for future action. 19/

17/ Report of Expert Consultative on Mediterranean Marine Parks and Wetlands, Tunis, (mimeographed), January 1977; see also Activities of the United Nations Environment Programme for the Protection and Development of the Mediterranean Region, October 1976, p. 3.

18/ "Phenomènes d'utilisation et de dégradation et instruments de sauvegarde des littoraux Italiennes" Rapport au projet concernant les bandes côtières méditerranéennes lancé par l'OCDE, Ministero del Bilancio e della programmazione economica, Instituto de studi per la Programmazione Economica, 1974, 324 pp.

19 L'Organisation Territorrale des Bandes Côtières Italiennes, ibid., 207 pp. with maps.

PART II

SECTION THREE

Legislative Policy and Regulatory Guidelines

CHAPTER 16

Legislative and Policy Guidelines

Coastal management policy or legislation at the national level should be viewed as means to an end and should thus be consistent with and form an integral part of a country's overall strategy for the achievement of national goals.

This is particularly true for smaller countries where coastal planning should be considered in the context of a national planning effort. For many island countries, that the coastal area is an economic region is very obvious when one examines its potential for growth and development, especially their recreational and touristic activities which are extremely attractive as sources of foreign exchange.

This chapter identifies some of the basic issues that should be considered in connexion with the establishment of such a policy or legislation, namely its objectives, acceptable level of uses, intensity of use, administrative set-up and means of implementation.

1. Establishing Objectives

The first group of issues relates to the establishment of basic objectives. Although management programmes may vary from case to case, the basic objectives usually fall within the three broad categories of development, conservation, or a combination of the two. The third group seems to cover a broad spectrum of cases since even in an overall development programme, there must be some elements of conservation; conversely, certain building or physical improvements are almost inevitable in a basically conservation-oriented programme. This is particularly true in dealing with a long stretch of coast or with areas involving divergent situations and interests which often require different treatments. Although basic objectives have to be implemented by specific measures, the establishment of the former determines the latter. Thus, for example, different measures may be taken to promote tourism and recreation on the coast. The appropriate measures depend on the basic objective. The British Countryside Commission, having adopted a conservation policy for coasts of high-quality scenery, decided that in seeking to encourage and promote forms of recreation, it was better not to introduce man-made attractions such as amusements and entertainment, or to provide large-scale or sophisticated recreation facilities of the kind normally found in a resort. The best measure was to install facilities that would enhance the enjoyment of the natural qualities of the coast. Another example is the coast of Scotland where the overall goal is the development of North Sea oil. 1/ But this does not exclude the establishment of conservation zones in areas of particular scenic, environmental or ecological importance in which major new oil and gas-related developments would in general be inappropriate, and could be justified only in exceptional circumstances.

The objectives should be related to specific and real problems. To illustrate, if the problem lies in the need for wiser use of natural resources, focussing on onshore and offshore resources within a coastal belt, the objectives should also be concerned with related issues, such as how to bring all resource agencies together in a co-ordinated machinery and which resources to include and exclude within such a system. This also raises the question of what should be the appropriate size of a management unit. In view of the extended resource jurisdiction of coastal states, it is also relevant to take into account the resources within the economic and fishery zones. From this resource-related perspective, a coastal area management objective may be regarded as an instrument of ocean management.

If the specific problem is environmental protection, the objectives may be conceived as means of facilitating the protection and preservation of relatively small and precisely delineated coastal areas of environmentally critical significance, such as saltmarshes, sand dunes, and estuaries.

Still, if the critical issue is to reduce the adverse effects of congested urban settlement, the objectives may have to be presented in terms of effective land use regulations, pollution control and industrial siting.

2. Classification of Coasts According to Acceptable Level of Use

In order to accord the proper treatment on the basis of resource capacity, some rudimentary form of zoning should be considered. A general measure is to classify coasts according to the acceptable level of use (i.e. their natural suitability for development and use) for each part of the coast involved. Three categories may be distinguished: firstly, those parts of the coastal area more appropriate for intensive development; secondly, those areas not suitable for development; and thirdly, all other parts which fall between those two categories and are suitable for controlled development.

The factors which may be used in the classification of these categories include geological considerations, ecological significance of the area and its tolerance to alteration, soil suitability, water quality, susceptability to flooding, archaeological and historical significance, environmental features and existing uses.

(a) Areas well-suited for intensive development are those which have elevations, soils, topography, and other physical conditions favourable to development and are not considered to be environmentally fragile. This category may also include lands having some physical limitation but suitable for development with certain modifications, such as an improvement in drainage or water supply. Control should be exercised in any event over all forms of development with respect to nature of activities, scale, siting, design, noise, waste and disturbance. Whatever form the management of a coastal area

---

1/ See Chapter 10.

may take, it is unlikely to be effective unless it is allied closely to a firm policy of land use and planning in general. Within each specific area, permissible and non-permissible uses should be determined.

There seems to exist a problem in some countries like New Zealand and Venezuela where too much coastal land is committed to urban development in a haphazard fashion and ahead of genuine demand. One way to deal with this problem is to select a reasonable number of localities along the coasts that should be developed for urban settlement in the next decades on the basis of such criteria as public preference, economic demands, qualities of the costline in the region, and local expectations. The resulting pattern of urban and rural land use can then be translated into reality by, for instance, district or local planning schemes, backed by and integrated with government programmes for public services.

Special attention should be given to creating urban development of an appropriate pattern. The task is to bring about forms of urban development which make the most of the recreational and aesthetic assets of each site. The standard of road access provided to any coastal area should reflect the function that area is intended to serve. Land use zoning in district schemes could be used in certain circumstances to modify the typical lineal form of coastal subdivisions and encourage development in depth; land use zoning could also be used to keep intensive urban development back from the beach. Land which is generally vulnerable to erosion should be protected by restrictive rural zoning to keep it largely free of subdivision and building. In addition to the use in district schemes of special ordinances tailored to the needs of coastal settlements, local government should also make clear through other means their policies for coastal development in general and for each settlement in particular. Local government should make a deliberate effort to explain their planning proposals to the public, and to stimulate the public to contribute their own ideas.

(b) <u>Areas which have overriding ecological, hydrological, physiographic or historical importance</u> to the public at large should not be committed to development projects which are incompatible with such natural conditions. Examples of these areas are submerged grassy areas essential to the propagation and nourishment of fisheries; low coastal areas covered by grassy, salt-tolerant vegetation subject to tidal ebb and flow during any part of the tidal cycle; selected estuarine beaches suitable for shore recreation with appropriate public access; areas of outstanding historical or archaeological significance. These areas are appropriate for inclusion in nature reserves and parks. There should be strict regulations regarding construction, roads and utilities, which should be limited to the minimum so as not to generate adverse effects on the environment.

(c) <u>Areas which fall between the above two categories</u> are those areas of the coastal zone that are not absolutely critical to regional ecological integrity, but because of their physical character, provide 'buffer zones' and represent retention-of-use options for future generations. This category may include aquatic preserves, aquaculture areas, river floodplains, scenic vistas, marginal lands, forest and game management areas, wildlife refuges. Some of these areas require special precautions when being converted to development in order to avoid undesirable consequences or harmful effects to the public's health, safety and welfare. For example, river floodplains and marginal lands (lands which owing to soil characteristics and drainage problems require major alteration in order to be made suitable for urban development) are appropriate for timber management, recreation and wildlife habitat. They are not suitable for housing or location of power generation. In these areas, conservation policy should prevail and schemes should be initiated to improve and enhance the appearance of landscapes by means of restoration, landscaping, tree planting and the removal of disfigurements. In these areas, opportunities for recreation should be explored to permit greater use to be made of the existing resources (such as antiquities, viewing points and other features of interest) without the need for any substantial increase in man-made facilites. This could be achieved by means of information services and a wide variety of language interpretation techniques to help the visitor's understanding of any features that may be of architectural, archaeological and historic interest.

3. Determination of Intensity of Use

The policy of management and the scale on which facilities are provided in a specific area of the coast should be related directly to the acceptable level of use determined. The concept of the intensity of use should be introduced for specific areas and be measured according to the maximum number of persons, kinds of activity, the pattern of access and the scale and provision of supporting facilities that the coastal area is able to accommodate without serious damage to its natural carrying capacity. In a conservation programme, for example, the carrying capacity of a resource may be considered in two distinct ways. Firstly, there is the ecological effect on the environment arising from the kind of the activities and the number of people involved. Above a certain level, the habitat will suffer as a result of discharge of wastes, trampling, disturbance to wild life, pollution, erosion of sand dunes. Secondly, there is the visual or environmental effect of people on the landscape; too many visitors will, as a result of excessive noise, overcrowding, and car-parking, destroy the character of the area to which they came to enjoy. Similarly, for an industrial project, it is necessary to take into account the location of development activities (on the coast or at some distance away from the coast), whether the land and water can absorb the development, existing communications and infrastructure, or the potential for improving them economically, and whether the environment can be rejuvenated, given the intensity of the intended use. Some of these issues are further discussed in the chapter dealing with industrial siting. 2/

Access by pedestrians and vehicles in both areas (b) and (c) mentioned earlier 3/ should be regulated in such a manner as to avoid excessive concentrations of these at points which are most easily damaged, or where the unspoilt or remote character of the coast or its scientific interest and importance is threatened. Road capacities and widths should be related to the acceptable capacities and levels of use of the resources to which they give access. Careful control of vehicular access and the management of traffic are basic to any conservation plan.

Policy with regard to anchorages, jetties and landing planes should ensure that the degree of access which may be possible from the sea is consistent with the management policy for the zone in question. Local authorities would need to consider carefully which means and what type of enactment to adopt in order to achieve effective management. Sometimes bye-laws may be desirable to confine direct access to selected points and to specify the type of vessel, permitted noise level etc. For many of these purposes, statutory powers may be needed.

2/ See Chapter 17.

3/ Subsection 2, (b) and (c) above.

Special attention should be given to certain use activities which are associated with environmental effects. For example, breakwaters may be needed to protect beaches, bluffs, dunes or harbour areas from wave action or for navigational purposes. But most of the solid breakwaters also obstruct the free flow of sand along the coast and starve the downstream beaches. Similarly, bulkheads and seawalls may protect the uplands, they do not protect the adjacent beaches, and in many areas are actually detrimental to the beaches by speeding up the erosion of the sand in front of the structures. It should be ensured that these structures are designed in such a manner as to avoid potentially detrimental effects on the movement of sand and circultion of water.

The removal of sand and gravel from coastal areas is a further example and usually results in erosion of land and silting of water. 4/ These operations can kill living marine resources. The removal of sand from the ocean beach can deplete a limited resource which may not be restored through natural processes. When rock, gravel, sand and minerals are removed from coastal areas, adequate protection against sediment and silt production should be provided.

4. Scope of Application of the Management Area

The seaward and landward limits of the scope of application of a management programme should be clearly defined.

It is extremely difficult to propose specific limits without referring to a concrete situation or studying the relevant factors. In this connexion, reference may be made to the experience of the Economic Commission for Europe in 1974 when it conducted a study on pollution of coastal and estuarial waters. 5/ An attempt was made for that study to propose physical boundaries for the purpose of examining the question of coastal pollution. The wide range of limits proposed by member states of the Commission indicated that such an attempt was not really feasible and this conclusion seems to be also applicable in the present case. Indeed, governments have adopted limits of very differing width for management purposes on the basis of their physical conditions and perceived needs. Some are broader than others; some are defined in a functional manner and others in specific figures: for example, Cyprus finds it necessary to include a seaward area 50 m from the high-water mark; Mauritius chooses the area from the coral reef to 1 km of the shoreline from the high-water mark; Sri Lanka, the strip lying between the limits of 3 m landward and 2 km seaward from the mean sea level; Togo, from the high tide within a landward area 100 m; the United States, from the water limit of its territorial sea landward to the shoreline to the extent necessary to control shorelands, the use of which has a direct and significant impact on the coastal water; Norway, all land areas where the sea and shoreline retain their importance for the location of vacation and tourist development.

4/ A case study on coastal erosion resulting from natural processes and man-made events may be found in Rapport, Seminaire-Atelier sur les problèmes de l'érosion des Côtes du Togo et de la République Populaire du Benin, organisé conjointement par les Nations Unies et les Gouvernements du Togo et du Benin, Lomé du 29 janvier au 9 février 1979, mimeographed, 55 pp., particularly Deuxième partie. 2.2.2.2.1 et seq.

5/ The Pollution of Coastal and Estuarial Waters, Economic Commission for Europe, Committee on Water Problems, ECE/WATER/6, 8 August 1974.

The present trend seems to be that a narrow strip of coastal area is preferred. Be that as it may, in order to permit effective management and to take into account any significant impact on the shorelands and water and on other economic activities in the coastal zone, the management area at national level should not a priori be limited to a narrow strip of the shoreland and water, particularly in view of the general development in many countries toward extending broad coastal resource jurisdictions. Perhaps the management area should include the area where resource jurisdiction is to be exercised.

For the purpose of management, it appears that the size of a given management area should take into account such considerations as nature of the activities involved, distance and locality of activities, physical impact of activities on other related activities, and the destination of the resources and materials produced (i.e. whether they are for export or for inland consumption). Thus, for example, the management area needed for oil exploitation on the continental shelf should cover not only the area where exploitation is taking place but also the immediate localities from which manpower and foodstuff come, in view of possible social and economic consequences. The size of a management area whose principal economic activity is sand and gravel mining can be, in principle, considerably smaller so long as the physical and biological impact areas are included.

5. Adequate Administration

The best policy and legislation will fail if the administrative machinery is badly organized or operated. For example, if the administration is entrusted to one group representing only one kind of interest, it is clear that the legislation will serve only that particular interest. Inefficient implementation of the law on the part of the administration may be due to various factors 6/ such as lack of co-ordination, lack of well-defined machinery to ensure a representation of different interests, shortage of staff or of sufficiently-trained staff, lack of proper equipment, and lack of awareness of the objectives of the legislation.

Proper administration should include (i) an administrative structure responsible for defining a coastal area policy and controlling the area, (ii) a systematic concept of administration, embracing all the activities with which an administration is concerned, (iii) the adoption of a defined coastal area as the unit for a management programme without prejudice, if possible, to the existence of political and geographical management units (economic development units), (iv) the avoidance of a prolifereation of new administrative institutions, and (v) interagency and intersectoral co-ordination and co-operation.

Experience shows that a well-representated policy-making body is most essential. The membership of such a body should consist of representatives of various government agencies (resource, planning, transport, economy), major interest groups (developmentalist, conservationist, and policy-makers) and qualified personnel to give technical advice. The danger is that a commission may become too large to work efficiently. To avoid paralysis, the chairman

---

6/ Among coastal management programmes in the United States, most states have found the question of appropriate administrative structure difficult to cope with. Further kinds of issue existed in the United States, and for the alternatives used in California, see H. L. Boschken, "Interorganizational Considerations in Coastal Management: the 1976 California Legislative Experience", Coastal Zone Management Journal, Vol. 4, nos. 1 and 2, 1978, pp. 47-64.

should be given a decisive role should a deadlock be reached. Weighted voting should be considered in view of the multi-sectoral interests of the coastal area and the need to weight different interests. Each interest group may be assigned appropriate voting power in accordance with a scale of importance of the interest it represents. In this manner, the decisions are more likely to reflect the harmonization of divergent interests.

It is also desirable for the administrative branch, particularly at the technical level, to be given explicit authority, within a well-defined framework, to translate into regulations the general norms. This system promotes flexibility, facilitates the implementation of legislation and makes allowance for regional differences.

The participation of interest groups and the general public in the planning and legislative processes, as well as in certain aspects of coastal administration, should be recognized and should be afforded legal guarantee and means of representation. This participation can be accorded both to individuals and to organizations representing a group's interests. The right of participation should enable different groups to put forward their concerns and needs where these are liable to be endangered by the various procedures established by the law in coastal areas. The representation of group interests can be assured by means of public hearings at the policy decision-making stage, or in the administration system, in which case an administrative organ or a tribunal is needed to harmonize conflicting interests.

Public participation requires education and training. If it has not been practised, it has to be cultivated.

6. Legislative Approaches

Although fairly comprehensive legislation on coastal area management already exists in several countries, these laws cannot be transplanted without careful adaptation. Nor should they be applied in isolation of policy objectives since they were based on particular physical and economic conditions. The legislation must reflect the perceived need of the region or area concerned.

The United States Coastal Zone Management Act, for example, is a framework legislation at the national level which leaves the states concerned to articulate their needs and define the methods to meet them. Scotland, Israel, Indonesia, Sri Lanka and Spain selected certain critical areas varying in size, and prescribed the necessary remedy. Other countries such as Venezuela, Mexico, Norway and New Zealand introduced special legislation to deal with certain critical issues, e.g. tourism, beach protection, or wetland management. 7/ There is no single solution and the need determines the legislative measure.

Before any drafting can be done, an inventory should be taken of all the activities concerned, all the legal provisions applicable to the area, and all the related administrative set-ups. After the inventory, the relevant laws should be classified and evaluated in the light of the basic objectives to be achieved. The examination should establish the adequacy of existing laws and of the administrative set-up vis-à-vis the achievement of the basic objectives, and should identify conflicts and gaps which may exist. After this stage, a decision on how to proceed further has to be made. Two basic alternatives are

7/ For details, see Chapters 10-14.

possible, to draw up self-contained coastal management legislation which would supersede and replace all existing applicable laws in the area so far as resource management is concerned, or alternatively, to amend existing laws by filling the gaps, removing conflicts, consolidating laws and co-ordinating administration. The end result of either approach is the same, that is to provide an adequate and coherent legal basis for the implementation of the objectives. A decision can be made only after weighing all the pros and cons in the light of the existing legal system and of the prevalent physical, economic and social conditions in the area concerned. Whether any of these alternatives is politically possible must be judged within the political and administrative set-up of a given country, and this issue is beyond the scope of the present study.

At the national level, it might be advisable to establish a general legal framework for coastal management within which the policy-makers, planners and executors must work. The law may determine the type and variety of information to be gathered, the procedures to be followed and the sectors of opinion which should be heard in order to legitimize the making of the policy and its implementation. The law may articulate broad principles, leaving the concrete policies and approaches to be formulated through the well-defined legal framework so as to reflect diverse economic and ecological conditions. While a general legal framework would thus be provided, regional (or state) and local governments would retain sufficient flexibility to formulate policies, to determine strategies, and to set out steps for implementation in the light of local needs and conditions. The drawback is that neighbouring local governments may come up with objectives conflicting with each other. A further step toward harmonization would therefore be necessary.

CHAPTER 17

Regulatory Guidelines for Industrial Siting

The question of the siting of industrial facilities on the coast (power generation, processing plants or factories) is seldom a feature in legislation dealing with industrial development. For many developing countries, government agencies involved are mainly concerned with fostering industrial development and they may be expected to show greater concern with economic goals than with other competing factors or values.

While coastal areas are ideal locations for industrial activities for obvious reasons of access, supply, market and waste disposal into the sea, the special characteristics of the coastal environment may constitute drawbacks which must be fully taken into account. Otherwise, disregard of these characteristics could result in the undermining of the very objectives intended to be achieved. Three categories of problems are involved. The first is protection against natural hazards (such as floods, storms, waves and current) which may have a direct impact on a given type of activities in the coast. For example, ignoring the existence of strong and direct wave action may render recreation on a beach extremely dangerous and in turn, would affect tourism and hotels located in that area. Thus, the development of a resort area would require studies of wave and current conditions to analyse their strength, direction and location, as well as their variability in order to identify areas along the beach where recreation is relatively safe. Otherwise, subsequent corrective measures such as the installation of breakwaters or groins would not only be costly but also would probably spoil the natural setting which made the location in question attractive for resort development in the first place. 1/ Similarly, hotels placed in areas where natural processes of erosion take place may in due course be endangered. Construction and building without knowledge of flood-hazard is another example of danger that may lead to destruction. In Lomé, Togo for example,part

1/ Several very interesting examples of this nature are cited in L. Neuman, "Interactions and Conflicts in Coastal Areas" in K. H. Szekielda (Ed.). Development and Management of Resources of Coastal Areas, joint publication by German Foundation for International Development and the United Nations, 1977, pp. 447-450. See also papers submitted on Benin and Togo which reported on problems of this kind in those countries, ibid., pp. 63-80; pp 235-242.

of the Hotel Tropicana has been eroded and the building itself could also be in danger of being eroded, if erosion continues at the present rate. 2/

The second category of problems relates to the complicated interaction of activities in the coastal area. 3/ For example, in Venezuela, population pressure on certain parts of the coast and the ensuing disposal of untreated waste water and physical encroachment led to the destruction of mangrove swamps, with a resultant loss of plant and animal life and possible increased erosion of the coast.

Estuaries and lagoons are areas of major conflicts of interest between fisheries and industry. The location and activities of industry may often interfere with the needs of nursery or breeding grounds. Similarly, industrial activities on the coast may also conflict with tourism and recreation, and all these activities increase the problem of waste disposal. Unless adequate means are found and activities carefully regulated, natural amenities may be reduced.

The third category of problems relates to the nature of certain activities. Nuclear power production, refining and processing and oil storage are activities having significant environmental implications. The siting of these activities onshore or offshore requires a basic knowledge of climate, winds, precipitation, storm frequency, currents, wave heights and directions, sediment transport, topography and landforms, and soil types. Experience shows that proper siting of industries can often ensure better development, avoid costly alteration and reduce adverse effects on other values, e.g. scenic, water quality, conservation of wetlands. The purpose of this chapter is therefore to identify the various issues that should be considered in relation to certain hazardous areas, marinas and offshore structures, and refineries and nuclear power plants. One common, yet important issue which is shared by all is the growth of licensing. This is dealt with in the last section.

1. Siting of Coast-dependent Activities

Fishing, aquaculture, port facilities, extraction of nearshore minerals (iron ore, salt, tin, sands and oil), tanker terminals, shipyards, marinas and certain applications of power generation may be referred to as coastal-dependent activities, in the sense that they must be carried out in or near the ocean to be able to function at all. Such industries, while obviously essential to a state's economy, can have major impacts on the coast, consuming valuable lands, intruding on the visual qualities of the coast, interfering with access and affecting air and water quality. Locations for such industries must take into account these effects, as well as safety concerns and growth needs. 4/

2/ See Rapport, Seminaire-Atelier sur les problèmes de l'érosion des côtes du Togo et de la République Populaire du Benin, organisé conjointement par les Nations Unies et les Gouvernements du Togo et du Benin, 29 janvier au 9 fevrier 1979, Deuxième partie, to be published in European Applied Research Reports, Harwood Academic Publishers, in 1981.

3/ Neuman, op cit., p. 450.

4/ The ecology of an island appears to justify special treatment. Some very useful guidelines may be found in J. McEachern and E. L. Towle, Ecological Guidelines for Island Development, published by International Union for Conservation of Nature and Natural Resources (IUCN), Morges, Switzerland, 1974, IUCN Publications New Series no. 29, 66 pp.

Legislation dealing with such industries should contain a clear statement of policy and criteria. A recommended policy would appear to be one that requires priority to be given to cases where the very nature of the development activity requires a coastal site, over other development activities on or near the shoreline. Where coastal-dependent industrial, commercial, and recreational development would have a substantial adverse effect on coastal resources, they should be regulated and permitted only if, for example, (i) there is no alternative site, (ii) economic benefits outweigh other considerations, e.g. the need for a port is essential to the region which otherwise possesses no economic resources, (iii) environmental damage is mitigated to the maximum extent technically and economically feasible in the design and execution of the project.

Coast-dependent activities should be grouped spatially, if possible. New developments should as a rule be concentrated in already developed areas unless for other economic, social or safety reasons (resources allocation and supply), such a concentration would not be desirable. All potentially hazardous industrial activities or other development activities that cannot be allocated or accommodated in already developed areas according to coastal area planning policies should be sited elsewhere and should be kept a safe distance away from population centres. All potential industrial sites in such areas should be used to the maximum extent feasible (subject to certain limitations such as safety requirements and economic necessity) prior to the commitment of any new areas. Sewage treatment, water purification, desalination, and power plant should be located when they do not interfere with and are compatible with recreational, residential or other public uses of the water and adjacent land areas. Waste treatment ponds for water-related industry should occupy as little shoreline as possible.

For many developing countries, economic and social values may sometimes attract more attention than environmental considerations; and thus, it is not always practical or economical to exclude non-coastal dependent industrial activities from the coast. For reasons already mentioned, siting regulations should therefore be considered for all major activities to be situated on the coast and likely to have a significant impact upon the environment, and upon social and economic conditions. 5/ Some of the attributes which may require scrutiny include the nature or type of activity, area of land use, location of proposed site, and number of employees. The relevant legislation should establish criteria on the basis of these elements.

Certain types of enterprise might be identified which, by their very nature, are more likely than others to create or result in significant or substantial impacts upon environmental, social, economic or physical conditions. Thus, energy generating and conversion activities, electricity generation, synthetic gas and petroleum production and uranium enrichment fall within this group. Other activities such as paper manufacturing, cement production and quarrying, livestock feeding on a large-scale, mining and mineral processing, chemical manufacturing, residential subdividing exceeding a specified number of units, textile milling, and tanning may also require siting regulations.

5/ R. F. Dasmann, J. P. Milton and P. H. Freeman, Ecological Principles for Economic Development, London and New York, John Wiley, 1973, provides a very stimulating discussion on these issues.

The area of land to be used in connexion with the proposed activity will often impinge upon environmental and cultural values as well as on social or economic conditions. The utilization of relatively large land areas may have adverse effects upon flora and fauna, scenic and recreational considerations and other natural systems. It is generally true, on the other hand, that the smaller the area is involved, the less the environmental effects will be.

Industrial development in areas exhibiting special cultural, scenic or recreational values, or areas peculiarly suited for wildlife habitats (such as wetlands, marshes, wilderness areas, rivers and streams) requires special scrutiny. The number of employees is another useful indicator in view of the necessary implications for transport, residence, communication, and waste disposal, etc.

Utility services, i.e. electric power, gas, communication, sewage and oil, are essential to coastal area development. The installation of pipes and wires must necessarily disturb the landscape, though it can usually be planned to have minimal visual and physical effects. 6/ Whenever utilities must be placed in a shoreline area, the location should be chosen so as not to destroy or obstruct the scenic view. If feasible, these facilities should be placed underground, or designed to do minimal damage to the aesthetic qualities of the area. Upon completion of the installation or maintenance projects on shorelines, banks should be restored, if possible to pre-project configuration.

2. Siting in Flood or Geologic Hazard Areas

Certain coastal areas are hazardous for development because of flood potential or other geologic reasons, and this should be taken into account in locating development. Activities in these areas should be specifically regulated by law.

The loss of life and property damage caused by flood are often due in large part to poor planning and the use of floodplains. Clearing of vegetation from the surface area can contribute to the intensity of flooding. The cumulative effect of many small structures is to reduce the floodplain's storage capacity. Along with changes in hydrologic characteristics of the watershed, such a reduction may increase the velocity of flood waters, thereby diminishing seepage necessary for groundwater recharge. Flooding may destroy valuable habitat areas and kill wildlife. In many developed countries emphasis has been on flood control projects which are very costly. The better policy seems to be to leave undeveloped many floodplains, where costly protection structures would otherwise be required. Substantial public funds can be saved by early planning.

Legislation can help by requiring careful studies beforehand and serious planning for development in flood hazard areas to avoid the need for new flood control works and to minimize the risk of interference with natural watershed processes, where such interference would adversely affect sand supply and anadromous fisheries. Criteria should be established in law for any new development in flood hazard areas. For example, allowable uses should be those which can sustain periodic flooding, and those which do not contribute to flood hazard, do not aggravate flood problems, impede floodwater storage capacity or increase pressure of existing flood control devices. Flood hazard

6/ See J. C. Sorensen, A Framework for Identification and Control of Resource Degradation and Conflict in the Multiple Use of the Coastal Zone, Department of Landscape Architecture, University of California, Berkeley, California, 1971.

areas should not be used during the flood-prone periods, for instance, for log decks or storage of materials that can be carried downstream by flood waters unless anchoring devices are adequately provided. Procedures should also be established to ensure review of proposed developments in inland flood hazard areas which could adversely affect lives and property in the coastal area.

Earthquakes, seismic sea waves and storm waves, landslides and mudflows, and bluff and shoreline erosion are the four major geologic hazards in coastal areas. Development that interferes with or ignores these nature processes may impose direct or indirect danger and costs on the public, and may accelerate or aggravate long-term natural geologic processes of the coast. Of direct concern for shoreline management are shoreline erosion processes. All areas of known geologic hazards should be identified and mapped. All proposed structures for human occupancy and other development activities which could significantly alter geologic processes or contribute to hazards should be carefully reviewed and regulated. Villages within the 100 year seismic sea-wave run-up zone should include within their safety elements a disaster preparedness plan which should include evacuation routes and an effective emergency warning system capable of adequately informing all residents and visitors of an impending occurrence. All these elements should feature in the legislation as requirements before a permit for development is given.

3. Regulations of Marinas and Certain Offshore Structures

Ports and marinas are essential for waterborne traffic and as such have become gravitational points for industrial and manufacturing plants. A coastal zone plan should recognize the regional nature of port services and should not permit development activities which have not been carefully planned. Since industrial piers and required water-areas adjacent to ports are usually longer and greater in bulk than those for recreational or residential use, the applicable law must require careful planning to reduce any adverse impact of such facilities on other water-dependent uses and shoreline resources.

The location and construction of offshore structures such as piers, jetties, groins, bulkheads or breakwaters should be regulated in view of their physical impact. Floating piers may constitute an impediment to boat traffic and can also alter beach sand movement patterns in areas where tides, longshore currents, and littoral drift are significant. Jetties employed at coastal inlets to maintain a navigation channel may often lead to the impounding of sand at the updrift jetty and to reducing the supply of sand to the shore downdrift from the inlet. The result is erosion. 7/ Similar effects may also result from groins, 8/ bulkheads, 9/ and breakwaters. 10/

7/ The construction of jetties in Lomé, Togo, and Cotonou, Benin, are vivid examples. For a detailed analysis of the causes and possible technical solutions, see Rapport, Seminaire-Atelier sur les problèmes de l'érosion des côtes du Togo et de la République Populaire du Benin, 1979, op. cit., Deuxième et Troisième Parties.

8/ Groins are barrier-type structures extending from the backshore seaward across the beach.

9/ Bulkheads are structures erected parallel to and near the high-water mark to protect adjacent upland from the action of waves or currents.

10/ Breakwaters are built offshore to protect beaches, bluffs, dunes or harbour areas from wave action.

The law should require that they be constructed in a manner that minimizes alterations of the natural shoreline, and does not result in adverse effects on nearby beaches. Sand movement and the effect of these structures on sand movement should be particularly considered in reviewing proposals for these structures. Their impact on wildlife propagation and movement, and the extent to which their design might detract from the aesthetic quality of the shore should also be reviewed.

4. Siting of Refineries and Nuclear Power Plants

Coastal areas are the preferred location for refineries and nuclear power plants primarily because of the free, abundant, and non-depletable waters of the ocean. But because of the very nature of these activities, which require large water supplies for cooling, oil terminals for transhipment and treatment facilities to handle waste, and because of the characteristics of coastal areas, special regulations are required. Adequate planning and siting considerations are particularly important. However, better knowledge of those factors deserving consideration is still required.

Refineries are large-scale and can be visually intrusive. Even the most modern refineries may occasionally emit noise and odours, and they represent significant single sources of air pollutants. Refineries can encourage the nearby construction of petroleum-associated industries (petrochemicals, plastics) which can lead to rapid industrial growth and increased employment. Refineries have a large potential for fire and explosion. Optimal safety considerations require siting refineries away from seismic areas, and separating them from surrounding populations by a buffer zone. Water pollutants emitted by oil refineries should be identified and regulated. Modern water-treatment technologies can reduce these emissions, 11/ and sometimes they can be eliminated through cooling systems which dilute pollutants to meet discharge standards. Removal of pollutants from the air and from water discharges of refinery systems will result in accumulations of solid or semi-solid waste products, for which proper disposal must be provided.

Before permission is given to site on the coast, it should be established that no alternative inland site can be found and that it is economically beneficial to locate the refineries on the coast. It should also be ensured that the project is designed and sited to minimize adverse environmental effects and that there will be a sufficient buffer zone to minimize any impact on surrounding property. The siting of a refinery in a highly scenic area, or in or near environmentally sensitive areas should be avoided if possible. The location of offshore moorings, storage terminals, and dockside loading terminals should be very carefully selected to ensure minimal disturbance of the environment, and the necessary structures should be equally carefully designed with that end in view. As a general rule, the construction of installations should be avoided in areas of high natural biological productivity, such as estuaries, mangroves, coral reefs and oceanic upwells. In other areas, strict operational regulations, inspection and enforcement, coupled with contingency planning should be implemented to reduce the risk of oil leaks or spills.

11/ For example, in Sumgait on the Caspian Sea in the U.S.S.R., a new purification plant is regulated to deal with 200,000m^3 per day of effluent from chemical works and oil refineries, yielding polymers and oil residues which can be burnt, providing methane to be fed into the city heating-system.

The major concerns relating to nuclear-power plant siting on the coast include hazards which are increased because of the sea-land-air interface, effects on the nearshore marine environment due to thermal or chemical discharge from cooling systems, proximity of nuclear power plants to population centres on the coast and the adequacy of emergency evacuation planning, and the impact of plant and associated structures on scenic natural areas. The radioactive hazard potential of nuclear power plants requires that the utmost care be exercised to site them away from areas of seismic risk and from population concentration. The principal constraint factors on inland siting are the availability of adequate water for evaporative cooling towers and the need to dispose of water of high salinity concentrated by evaporation in the cooling tower. If sufficient cooling water is available, inland siting of nuclear power plants is to be preferred - at least it can reduce, if not eliminate completely, the possible effects on the sea. Because of the importance of fresh water for agriculture and other uses, however, inland siting raises other kind of problems.

5. The Granting of Permits

The permit system can be used as an effective method to control the siting of industry in coastal zones. Although a permit-issuing system exists in most countries, many existing systems do not specifically include the siting of industries. Some systems, particularly in industrialized countries, often require either diverse or numerous applications and permit proceedings before different and unrelated central, state or local government agencies. 12/ The procedures are often time consuming and sometimes may result in omission, confusion or inconsistency. Red tape may also discourage initiatives. It would seem that there should be one, co-ordinated proceeding before a co-ordinated system or preferably, a single agency empowered to issue or deny a siting permit covering all aspects that should be considered. This would not only ensure better regulation and control, but also benefit the applicant and other parties having an interest in the probable effects of the proposed facility.

A single or co-ordinated permit system for the siting of industries should include the following elements. Firstly, the licensing procedure should encompass all aspects of the proposed project so that, at its conclusion, the siting agency may issue or deny a single permit on the basis of recommendations received from all regulatory agencies which might otherwise have licensing or approval authority over any aspect of the proposed project, and also binding upon all other interested parties. The final action of this agency should be reviewable, at the instance of any party to the permit proceeding, in a single judicial review. Secondly, recognizing that the expertise of many governmental agencies may be essential to evaluate various aspects of a proposed project, the single proceeding should combine and co-ordinate the activities of all governmental agencies with respect to the project in order to avoid overlapping or fragmentary reviews which may be in some respects redundant or in others inadequate. Thirdly, in order to achieve an appropriate balancing of the desirable effects of the project (essentially those involving the benefits of industrial and commercial development) against its undesirable effects (those involving adverse impacts upon existing

---

12/ For example, a large chemical corporation gave up its plan to build a $500 m petrochemical complex in northern California, U.S.A. because of the costs and time required in connexion with obtaining the necessary 65 permits (five from the Federal Government, 40 from the State of California, and 20 from the local counties).

environmental, social and economic conditions), the single-permit agency should be authorized to balance the project's benefits against its possible detriments and, in reaching its permit decision, should be empowered to override decisions or recommendations of other governmental agencies which may retain jurisdiction over only one or a few aspects of the proposed project.

CHAPTER 18

Regulatory Aspects of Oil Pollution Control

The spilling of oil and other noxious substances in coastal areas may have not only harmful effects on the environment, damaging to natural beauty and wildlife, but also economic and social consequences on tourism and fishing. Oil pollution thus represents one of the most serious problems in coastal management. 1/

Certain coastal areas and islands conveniently located in relation to high-demand markets are likely to experience an increased risk of pollution due to a transformation now under way in the oil transport industry. This stems from the gradual replacement of traditional tankers by very large crude carriers having high draughts (90 ft). Islands located near markets surrounded by deep water offer obvious advantages for storage or refining before oil is shuttled in smaller tankers where and when required. New and expanded facilities have thus been proposed or scheduled, for example, in the Caribbean area (Jamaica, Haiti, Puerto Rico, the Caymans, St. Croix, Turks, the Bahamas and Guadeloupe). 2/ Oil pollution hence presents a special problem for these areas. Measures to control oil pollution should accordingly constitute an integral part of a coastal management programme.

The overwhelming majority of oil spills are caused by human error rather than by equipment failure. 3/ The risk of spills can never be entirely eliminated and there always exists a danger of an accident and oil spill in the neighbourhood of an oil installation and in areas where oil tankers regularly pass along the shoreline. Clean-ups may require special equipment which may not be locally available and can be very costly. Moreover, all this

---

1/ See Section II, Manual on Oil Pollution contained in IMCO document MEPC IX/9, 17 March 1978. Section II deals with contingency planning. Also see Part IV: Organization Aspects, in J. Wardley-Smith (ed.), The Control of Oil Pollution on the Sea and Inland Waters, London, Graham and Trotman Ltd., 1976; pp. 203-223; Accidents at Sea Causing Oil Pollution, Review of Contingency Measures, London, United Kingdom Department of Trade, 1978.

2/ See J. McEachern and E. Towle, Geological Guidelines for Island Development, published by International Union for Conservation of Nature and Natural Resources (IUCN), Morges, Switzerland, 1974, IUCN Publications New Series no. 30, 66 pp., p. 50.

3/ Manual on Oil Pollution, op. cit., p. 205

cannot be arranged after the threat of pollution is imminent. Unless appropriate contingency arrangements have been made beforehand, serious mistakes and irreparable harm are inevitable.

Some of the basic issues are discussed here, grouped within the following two sections.

1. Planning Guidelines

Effective oil pollition control must begin with proper planning which should take into account the probable size of a pollution accident, equipment needs, classification of coastline and location of exploration, storage and refining activities.

The risk and the probable size of an oil spill should be reviewed in the light of the following factors. 4/ the amount of oil being transported into or out of the area and being transported close to the shore concerned; the past record of the area for oil spills; the size of tankers which most commonly pass through the area; the likelihood of collision between tankers or between tankers and other kinds of vessles; the likelihood of grounding; the maximum rate of discharge from a well, refinery or terminal if the control devices prove effective; the length of time required before the flow can be checked or stopped altogether; and current and weather conditions. Once the risk and probable size of a spill are determined, it will then be possible to establish a target in relation to which contingency arrangements can be made. When an accident does take place, it can easily be ascertained whether the contingency arrangements alone will be sufficient or whether additional help will be needed. In the latter case, steps can be taken to obtain additional assistance.

When oil moves landward from the sea, equipment will be needed to keep the oil away from the coast (booms), 5/ to disperse the oil at sea before

---

4/ See Chapters 5 to 8 in J. Wardley-Smith (ed.), The Control of Oil Pollution on the Sea and Inland Waters, London, Graham and Trotman, Ltd., 1976.

5/ Oil booms are floating barriers supported by one or more buoyancy units used to prevent the spreading of oil on water and sometimes to thicken the layers of oil by reducing the area into which it has spread. Mechanical devices for the removal of oil from the surface of water are known collectively as "skimmers", which, according to experience, do not, probably cannot, achieve the intended purpose. Sometimes, other methods such as using absorbents or burning, gelling, or sinking may also be considered. The use of dispersants has been the most widely-employed and immediately effective method both on the open sea, in estuaries and on beaches. When oil floating on the surface of the sea is treated with a suitable surface active agent and an agitative is applied, the oil is broken up into particles of different size, some small and some large. Once the oil has been dispersed, the natural motion of the sea is usually sufficient to distribute the particles over a wide area and prevents their re-forming into a slick. The bacteria naturally present in the sea will then attack the small particles. The normal method of applying the dispersant to the oil is by means of spray booms fitted to ships. Experience shows that dispersants should be used according to the manufacturers' instructions and with suitable and efficient spraying equipment. It is not sufficient simply to have supplies of dispersant and equipment available at the right place at the right time; trained personnel must also be available. For a very useful examination of the various methods and equipment that may be used for the treatment of oil on the water, see J. Wardley-Smith (ed.), op. cit., particularly Chapters 5-8, pp. 83-159.

it reaches the coast, and to contain and pick up the oil. It is much easier to disperse oil while it is still at sea, and every effort should be made to do this before it washes ashore. Experience shows that it is much easier to deal with oil on the sea than on shore. 6/

Moving oil from the shoreland is seldom an easy matter and no single method or device can be recommended. 7/ The situation is complicated by the different physical characteristics of the oils (light, heavy, gasoline, or diesel) and the many different types of shoreline which can be anything from soft mud to hard rock. 8/ A variety of equipment may be needed: earth-moving equipment to remove dirty beach material, oil pick-up equipment, hand-held and other types of dispersant equipment, sea water hoses etc. Stocks of expendable material must be purchased and stored. The most obvious approach to beach cleaning is to remove the oil physically. This is far from easy on many beaches as the oil gets mixed with, and coats, the beach material. Sometimes, it may be necessary to burn the oil, cover it up with an absorbent powder or wash it away using a dispersant chemical. Each of these methods can be of value in particular circumstances. 9/

The decision to clean or remove oil on the shore is conditional upon two important considerations: whether the cleaning or removal is necessary and whether the cleaning process will damage the environment. On the other hand, the presence of oil effluents or lumps of tar (tar-balls), is unpleasant for tourists and from the amenity point of view. But the cleaning process often causes as much concern among biologists as the oil pollutants themselves, and in some cases cleaning may increase damange to shore life. 10/ Accordingly, a clean-up policy should be developed to give guidance on when and how to deal with spills, and under what circumstances these should be left to disperse and degrade by natural means.

6/ ibid., Chapter 8.

7/ ibid., Chapter 2.

8/ See below regarding the types of shoreline and their vulnerability.

9/ See for example, Ottway, S. and Knight, M., Review of World Oil Spillages. Marine Ecology and Oil Pollution, Applied Science Publishers, 1975. There is considerable published material concerning the effects of oils and dispersants on the plants and animals of the shore. The relevant factors which appear to be important in determining the extent of damage include those which are related to the oil and cleaning treatment (type of oil, type of cleaning treatment) and to biological and other natural circumstances (topography, hydrography and the climate of the affected area).

10/ It has been observed that whereas in many instances, there is much merit in doing nothing at all to the oil, in other instances, cleaning is called for. If, for example, the oil threatens amenities, then it should be removed. But if the oil spill occurs in winter on a fairly remote coastal area largely used by holiday-makers in summer, then, providing it will have been taken away by natural causes before the start of the holiday season, perhaps the clean-up should not be carried out.

Many issues must be considered in making a decision on the correct control and treatment of shore oil pollution, as for example the short and long-term effects of a particular type of oil, the efficiency of different cleaning methods and the damage their use causes to shore life, and plans for the discharge of ballast water or effluent from a terminal or refinery. Moreover, it should also be noted that the distribution and persistence of spilled oil may be influenced by a complex interaction of various factors 11/, such as wind stress, water currents, beach activity and sand-grain size, wave energy and oil quantity and composition. An effective control programme must not ignore the interaction of these and other factors.

If possible, the entire coastline, or at least those parts of the coastline where oil might pollute, should be examined in detail by personnel familiar with the district and qualified in wildlife, oceanography and related field to determine, in collaboration with those organizations dealing with conservation and/or tourism, which coastlines must be cleaned and which need not. All this should be done as part of the planning process. This task cannot be left until the oil is on the shore. For areas where cleaning is required, the precise type of cleaning permitted and the extent of the operation must also be decided in advance and included in maps prepared for that purpose.

It may readily be seen that a shore classification according to physical characteristics (sand, gravel, mangrove, etc.) and economic activity (tourism, industry, archaeology, etc.) can be very useful for achieving the tasks mentioned above. The technique used for parts of the Spanish coast illustrates this point. 12/ It has also been pointed out in a study that according to physical processes and shoreline characteristics, as well as biological effects and long-term oil persistence, different types of coast may be identified in terms of their vulnerability to oil spill damage. 13/ Thus, a shore classification can also help in the determination of vulnerability.

On the basis of oil movements, the whole coastline or those parts which might be polluted should be divided into administrative units for the purpose of oil pollution control. In each area, there should be one person designated as the controller who should be accessible at all times and is responsible for making necessary decisions. If the district has been classified as suggested and plans have been made in the event of an oil spill, the controller's task is to ensure the application of the appropriate plan according to a given

---

11/ E. R. Gundlach and M. O. Hayes, "Vulnerability of Coastal Environments to Oil Spill Impacts", Marine Technology Society Journal, in press. By the same authors, "Some Guidelines for Oil-spill Control in Coastal Environments: Based on Field Studies of Four Oil Spills". Paper presented to ASTM Symposium on Chemical Dispersant for the Control of Oil Spills, Williamsburg, Virginia, November 1977.

12/ See Chapter 13 regarding coastal planning in Spain.

13/ Gundlach et al., op. cit.: The following types of coasts are listed in order of their increasing degree of vulnerability: (i) Exposed, steeply-dipping or cliffed rocky headlands; (ii) eroding wave-cut platforms; (iii) flat, fine-grained beaches; (iv) steeper, medium- to coarse-grained beaches; (v) exposed, compacted tidal flats; (vi) mixed sand and gravel beaches; (vii) gravel beaches; (viii) sheltered rocky coasts; (ix) sheltered estuarine tidal flats; (x) sheltered estuarine salt marshes and mangrove coasts. (Coral reefs were tentatively placed between (vii)-(viii).

situation. All oil spills at sea should be reported and channelled to the controller in the area. Equipment for spraying dispersant should be availble throughout the coast. Prior arrangements should be made in every area so that in the event of a spill, ocean-going tugs from home ports can be fitted with equipment to go out and deal with the oil. The tug crews should be trained, and contingency drills should be carried out on a regular basis. Provision should also be made by the local government to deal with oil on the beach and shore.

2. The Role of Government and Administrative Requirements

In view of the potentially harmful effects of oil pollution on the environment and the scale of arrangements that may be required, it would appear that government should play a leading role to ensure that plans and resources, whether its own or of other organizations, exist to cover all areas of activity from which oil pollution may arise. Government must ensure that individual plans interlock to form a viable, effective and compatible whole, each agency being aware of the existence and spheres of responsibility of the other.

As already mentioned, 14/ in the case of the United Kingdom, the Government is responsible for making a contingency plan to deal with oil pollution and it actually carries out the clean-up. The United States' Government, on the other hand, is responsible for the preparation of a central contingency plan and plays a supervising and supporting role. The law sets out guidelines for what is required and requires the spiller to be responsible for the clean-up. When an incident occurs and when a federal agency is informed, the procedures of the National Oil and Hazardous Substances Pollution Contingency Plan are put into operation. 15/ An on-scene commander comes to active duty. The commander's task is to oversee that a satisfactory job has been done in accordance with what is required by law. Government intervenes only when the spiller is unable to assume the task, and then undertakes the work either directly or by employing a contractor. The United States' approach is very much related to the fact that there has existed for some time commercial firms or oil spill co-operatives (over 100) specializing in oil spill clean-up. In practice, most, if not all, of the clean-up operations are executed by those companies.

It appears that either of these approaches can serve as a model for contingency planning.

The main policy issue is the role and responsibility of government and of the owner of a refinery, terminal, barge or tanker causing an oil spill. So far as the clean-up is concerned, the responsibility depends very much on the policy that government has chosen. If the government is to require the spiller to carry out the clean-up, the law must clearly stipulate what is expected from the owner; how the clean-up is to be carried out and in what manner; who is to oversee and be responsible for implementation; what kind of supplementary measures may be needed and provided by the government itself if the spiller is unable or fails to carry out the task. It is a very difficult task to ensure that the owner will be adequately equipped or will have access to adequate clean-up facilities. If government itslef is to carry out the clean-up, the law must require the owner to carry adequate financial liability in order to reimburse the State in the event of an oil spill. To set the financial liability at the appropriate level is a difficult question. If it

14/ See Chapter 11.

15/ *ibid*.

is too high, the result could mean higher costs for the transportation of oil. If it is too low, the amount may not be sufficient to cover the clean-up.

The legislation may also need to deal with the supply of equipment, organization, and communication, and to enlist support from local government, the public and the relevant government agencies dealing with transport, trade, security, the coastguard and the environment. The law should be prepared to be consistent with the maritime jurisdiction of the state. In congested areas such as straits or areas of enclosed sea where neighbouring states may be involved, co-ordination may be necessary for effective implementation. In these and other cases, a regional approach might be desirable. The joint programme between Canada and the United States may serve as a model. 16/

There should exist appropriate arrangements for consultations by all anti-pollution agencies with organizations representing fisheries and wildlife interests, both during the planning and at the operational stages. Arrangements should exist for training in the relevant techniques for combating pollution, and the personnel concerned should take advantage of these facilities. It may be mentioned that advantage should also be taken of the various industry and government courses which already exist in some countries, as in the United States, Canada and the United Kingdom, for the training of those responsible for the handling of pollution incidents. These courses not only include guidance on types and location of potential spills and the various methods available for dealing with them, but also give practical instruction in the handling and maintenance of equipment.

A recent IMCO publication entitled Manual on Oil Pollution 17/ can usefully serve as a guide to making detailed arrangements for oil pollution control.

16/ See Chapter 15.

17/ Manual on Oil Pollution, particularly Section IV, IMCO publication, December 1978.

CHAPTER 19

Guidelines for Coastal Construction Setback Lines

1. Introduction

Coastal beach erosion is a common problem in many parts of the world. The causes of beach erosion are either natural, e.g. tidal and wave action or man-made, e.g. groins, jetties or seawalls. Sometimes, it is a combination of both. Coastal construction setback lines are surveyed lines that define the seaward limits of construction activity along the land portion of the coast. The location of these lines is based on scientific criterion regarding beach erosion patterns.

The purpose of the establishment of setback lines is two-fold. In the first place they are intended to ensure that beach erosion is not triggered or worsened as a result of coastal construction activity which interferes with beach processes. However, this may not always be possible or successful. For example, the location of a harbour, a seawall or an artificial inlet is often determined by economic or physical needs, rather than by the consideration of erosion. The existence of such lines does serve, however, as a warning. If a decision is made to carry out construction activity beyond the limits, consequences must be faced. In the second place, perhaps the primary objective is to protect building construction, e.g. housing, roads or plantations, by locating them away from the zone of probable worst-case erosion.

Setback lines are applicable in areas where erosion is already active, and also on presently stable beaches. They find their most important application on coastlines where important development is planned or foreseen. Coastal zone management planning for an area with multiple goals - industry, housing, recreational facilities, etc. - can conveniently be carried out after a setback line has been established. For example, suppose a 50 year coastal development plan is to be established for a given coastline. Plans call for the development of fishing piers, industrial zones, sewage treatment facilities, tourist and recreational facilities, fish and wildlife reserves. As a first step, a coastal setback line can be established, and 'areas of concern' determined from ecological considerations. Then the coastal area can be zoned appropriately, in order to accommodate the multiple uses. This is done in such a way that maximum exploitation can be carried out with the least impact on erosion and ecology.

Countries that border on a common coastal environment have a particular interest in the establishment of a regional coastal construction setback line. A good example is the area of the Gulf of Guinea, West Africa, bordered by Liberia, Ivory Coast, Ghana, Togo, Benin, Nigeria, Cameroon and Gabon. The coastline is characterized by an important longshore sand transport rate of about 1.5 million cubic yards per year in a west-to-east directtion. Development plans over the past 15 years in this region have included important dams on rivers (which effectively trap the sediment supply to the oceans), and harbours with shore-perpendicular jetties (which arrest the longshore sand transport, causing beach erosion west of the harbours, eventual silting in the harbours, and severe erosion to the east). In addition, expensive industrial and tourist complexes have been developed on beaches that later were subject to the influence of beach erosion. Regional setback lines drawn over this entire area would not only aid in national planning for each country, but would also ensure that coastal development in one country does not imperil the beaches in another country downdrift. In regions such as this one, each country is afforded protection from beach erosion when the adjacent country situated up-drift adopts coastal setback lines near the border between the two.

Setback lines are not primary corrective measures that can restore an eroded beach to its original configuration. Rather, they take into account the probable erosion that would occur under the worst oceanographic, geological, and meteorological circumstances, and define the seaward limits to which development may be safely carried out. The necessity of considering additonal corrective measures for beach erosion is reduced, if not totally eliminated, when the setback line is respected.

2. Causes of Erosion

Coastal erosion can be classified into two types, natural erosion and man-induced erosion. While the former is a relentless process which is often impossible to resist, the latter is often due to man's ill-planned activities and can certainly be contained, or even reversed.

(a) Natural causes of erosion. A natural shoreline is established as a result of the interaction between geological factors, fluvial and marine contributions of sediments, as well as meteorological and oceanographic conditions. Any variation in one of these factors will cause the shoreline to advance or retreat.

It is generally agreed by geologists that, due to deglaciation of the polar icecaps, the sea level has gradually risen about 100 m over the past 20,000 years.

It is also apparent that the rise has greatly slowed down over the past 5000 years. The present-day rise in sea level is estimated to be in the order of a few millimetres per year.

Tectonic adjustments in a given region can further increase the relative rise in sea level. Consequently, flooding can affect large coastal areas wherever the coastal plain is flat.

The combined action of waves and tides on the coastline affects the stability of the shoreline. If, for any reason, the sediment supply to a section of beach is reduced, the constant pounding of waves can cause severe erosion. This phenomenon occurs in many parts of the world. Even rocky cliffs can be undermined at their base and slump into the sea. For example, in the Pays de Caux, erosion has been 0.3 to 0.5 m per year, and 0.6 m per year along the Côte de Biarritz in France. 1/

River deltas are especially prone to dramatic changes. Tributaries are abandoned as the river shifts its course. The abandoned delta lobes, having lost their sediment supply, erode as a result of wave action. New deltas are created and prograde into the ocean, the sediment supply overcoming the destructive wave effects. In microtidal regions, these wave effects are very pronounced. 2/

The presence of offshore ridges can also be a cause of erosion. These ridges often change their position and size with time. Waves approaching from deep ocean are refracted as they pass over the ridges and converge on a section of the beach which then erodes.

Catastrophic events like tsunamis, severe storms, and hurricanes (cyclones, typhoons) cause the sea level to rise to abnormal heights. Severe erosion ensues.

Historical records of such events permit prediction of the probable catastrophic water uprush on the coastal area, and the resulting erosion. Comprehensive surveys of storm surges have been carried out for a number of areas, including Japan, the Florida coast of the United States, and Hong Kong. 3/ Although catastrophic events are generally infrequent, the potential damage is severe enough to warrant consideration of locating population centres and industries away from their area of influence.

(b) Man-induced erosion. Coastal activities can also directly or indirectly result in beach erosion. The following are some examples:

(i) Dams on rivers reduce or arrest the river-borne sediment supply to the ocean, upset the sediment budget and can, in some cases, deprive the beach of its normal source of sand.

(ii) Harbours often have shore-perpendicular solid quays which stop the longshore transport of sand and cause accretion on the updrift side, and erosion drowndrift.

(iii) Sand removal from the coast upsets the longshore sand transport budget and can result in erosion downdrift if carried out at significant levels.

(iv) Groins and jetties and other shore-perpendicular structures interfere with longshore sand transport and can result in erosion when used improperly.

1/ C. Migniot, "Problèmes d'érosion littorale", Recontres Internationales de l'Ouest Français, 11 May 1979, mimeographed (unpublished).

2/ See M. O. Hayes and T. W. Kana (eds.), Terrigenous Clastic Depositional Environments, Tech. Report, no. 11-CRD., Dept. of Geology, University of South Carolina, p. I-1 to 120.

3/ See R. Silvester, Coastal Engineering, Vol. 22, Amsterdam, Elsevier, 1974, Chapter 4.

(v) Artificial inlets, which may narrow waterways connecting bay or lagoon to the ocean, can often have high current velocities which prevent longshore sand transfer from one side to the other.

(vi) Seawalls, bulkheads and breakwaters, which are shoreline protective structures, have side-effects in terms of erosion of adjacent areas.

It is commonly recognized that once a coastal community is established in an area subject to beach erosion, a wide variety of protective structures may be called for or used in an effort to protect individual concerns. For example, just within the Buffalo District of Lake Erie and Ontario where severe erosion occurs, there are some 35-40,000 coastal structures. The combined effect of these structures is extremely difficult to assess in as much as each structure is itself likely to be a cause of erosion elsewhere.

3. Criteria for the Determination of Setback Lines

For the establishment of setback lines, a number of factors that allow prediction of the probable extent of erosion over a given period of time must be taken into account. These factors include the nature of the shoreline; winds, waves, and tides; historical storms; storm surge and wave set-up; beach and offshore profiles; wave uprush; ground elevation; erosion trends; vegetation line; dune line; and upland development.

(a) Nature of the shoreline. The type of bedrock, the curved or straight-line configuration of the beach, the presence of rocky headlands and of estuarine areas, and the type and size of beach sediments are some of the characteristics that affect the amount of erosion that can occur.

A hard granite or basalt bedrock tends to slow down or arrest erosion, while a soft clay tends to be eroded away; the alignment of the beach with respect to breaking waves affects the longshore sand transport which must be considered in erosion calculations; oblique waves striking a coast situated between two rocky headlands cause erosion of the shoreline until an equilibrium crenulate-shaped bay is formed; rocky beaches are harder to erode than sandy beaches.

The nature of the shoreline must be studied in order to determine the beach areas which are most prone to erosion and the short-term extent of erosion. Both geomorphological and physical considerations need to be applied.

(b) Winds, waves and tides. Winds, waves, and tides have a direct impact on coastline stability, especially on short-term trends and fluctuations of the beaches. Records of these parameters should be obtained for the area of study, or extrapolated from neighbouring stations. The sudden impact of severe weather can remove, in a few days, as much sand as is normally moved over a period of months or years. However, in some cases, 'normal' conditions can cause more erosion in the long run than short-lived severe conditions.

Statistical analysis of historical wind and wave records yields their predominant characteristics which affect beach processes. Additional measurements can be taken to supplement the available records. 'Spring tide' is the term used for astronomical tides with the highest vertical range between successive high and low tides. It occurs about every two weeks, at new or full moon. The erosion action of waves extends furthest inland at spring high tide. Tidal data are generally available at harbours, and can be extrapolated to the area of study. An automatic tide recorder installed at the study site will also provide control of site-specific data.

(c) Historical storms. Storms and hurricanes (typhoons, cyclones) can cause an abnormal rise in sea level. The interaction between wind and water occurs in different ways: reduced atmospheric pressure causes a sucking effect on the ocean waters; swell waves are generated which propagate faster than the storm itself, peak up and break in shallow water, thereby increasing the water level; the wind exerts a shear stress on the water surface, causing it to pile up in shallow coastal regions.

Examination of historical storm and hurricane records can yield a 'design' or 'standard' hurricane for use in the area of study.

(d) Storm surge and wave set-up. Storm surge (or wind set-up) is the vertical rise in still water near the coast, caused by storm wind stress alone (this is in addition to the astronomic tide). Storm surge can be calculated for the area of study if there exist enough data on high water levels; otherwise, it can be extrapolated from data from nearby regions. 4/ Storm surge can also be evaluated from known meteorological conditions. Different mathematical models of varying complexity can be used. 5/

Wave set-up is the superelevation of the water surface over normal surge elevation and is due to onshore mass transport of the water by wave action alone. Wave set-up is sometimes estimated as 1/20 of deepwater wave height. The highest wave set-up is expected where the continental shelf is narrowest, and offshore waves are highest. Empirical estimates of wave set-up, from measured wave characteristics, are discussed in existing literature.6/

(e) Beach and offshore profiles. The slope of beach from a point on the beach to a depth of a few metres below mean sea level will affect the wave uprush on the beach. Topographic profiles on the beach down to a wading depth can be done by standard surveying techniques. Hydrographic profiles in shallow water are obtained by the use of a small boat with an onboard echo-sounder. The topo- and hydro-profiles are then joined to produce continuous ground and offshore profiles for use in computing wave uprush. Offshore contours in deeper water can also be used to study the convergence of waves on certain areas of the beach with resulting erosion zones. Ground contours can be obtained from controlled, stereoscopic vertical aerial photographs of the coastline.

(f) Wave uprush. Wave uprush is the rush of water up the beach following the breaking of a wave. It depends on the wave characteristics, the slope of the beach, and the roughness of the beach material. Setback lines should be situated at a minimum distance beyond the wave uprush.

Calculation of wave uprush for each beach profile is usually done in the following manner. First, a worst-case stillwater level is obtained by addition of the storm surge, the wave set-up, and the spring high-tide level. Then, a representative storm wave of estimated wave height and wave period is chosen to compute the uprush. The computation can be done by the composite slope

---

4/ J. A. Purpura, Establishment of a Coastal Setback Line in Florida, in Proceedings of 13th Coastal Engineering Conference, Vancouver, Vol. II, 1972, pp. 1599-1615.

5/ For example, Shore Protection Manual, U.S. Army, Coastal Engineering Research Centre, Vol. 1 (Third Edition), 1977.

6/ For example, ibid, Chapter 3.

method 7/, if the beach slope is irregular. Laboratory tests can be carried out to determine the effect of beach roughness on the uprush. A computer programme can be conveniently developed to handle the tedious calculation.

(g) Ground elevation, erosion trends, vegetation line, dune line, and upland development. After having done the wave uprush calculations, consideration must be given to the following factors:

- areas of high ground elevation will generally be less exposed to flood and erosion;

- erosion trends as determined by analysis of aerial photos and from geomorphological considerations will yield rates of erosion and indicate areas of pronounced erosion;

- the vegetation line and types will give insight into the advance and retreat of the waterline over the past several decades;

- the longshore sand transport rate is examined to estimate the probable extent of erosion in critical areas;

- the dune system which takes part in the dynamic processes on a beach should be preserved seaward of the setback line so as to provide added protection against erosion.

4. Case examples.

Four examples are cited here to illustrate the approaches that can be taken regarding the establishment of coastal construction setback lines. The setback line drawn in Florida, U.S.A. in 1972 is illustrative of an approach suitable for sandy beaches exposed to storm and hurricane action. The setback line prepared for Kiawah Island, South Carolina, U.S.A. illustrates the geomorphological approach to beach processes on a barrier island. The artificial use of crenulate-shaped bays to stabilize the Singapore coastline is the third example. The fourth example, Charente Maritime in France, is intended to demonstrate how a setback line can be used to further the goals of coastal zone management.

(a) Coastal setback line in Florida 8/. The study areas consisted of a stretch of 35 km of beach on the Atlantic Ocean, characterized by a tidal inlet, several barrier islands, and by a history of erosion. A comprehensive field programme was carried out. Monumented stations were placed about 300 m apart, parallel to the shoreline. Each station was surveyed and tied to the Florida State Plane Co-ordinate System. Beach profiles were run from the back of the dune to a wading depth at each station. Offshore soundings from the beach to a depth of about 10 m were done at about 40 stations. An automatic tide recorder was operated continuously during the study. Stereoscopic aerial photos were taken. Ground studies were done to examine beach material and composition, rock strata, vegetation-bluff-scarp lines, coastal structures, and development.

7/ See T. Saville, Jr., Wave Run-up on Composite Slopes, in Proceedings of Coastal Engineering Conference, 1958, no. 6. Council of Wave Research, published by University of California, 1958.

8/ J. A. Purpura and W. M. Sengabangh, Coastal Construction Setback: Florida Co-operative Extension Service, Marine Advisory Programme, Rep. No. SUSF-SG-74-002, 1974.

Historical data were analysed to determine erosion trends, winds, waves, and tides. The net littoral sand drift was estimated. A storm surge frequency curve, obtained from a study of storm tides in the area, was developed. The curve yielded storm surges that are probable over desired time periods. Wave set-up for a 'standard hurricane' was estimated. A computer programme was developed to perform calculations of wave uprush on each profile by employing the composite slope method.

A storm surge of 4.2 ft (with probable occurrence of once in 20 years), 2.0 ft wave set-up, and 1.8 ft spring tide gave a combined stillwater level of 8.0 ft under storm conditions. Computation of wave uprush was done for a wave height of 6.5 ft and wave period of 8.0 sec. shows the 20 year probable level, the computed wave uprush on two beach profiles. The position of the setback line is indicated on each profile where it crosses at the point (SBL).

The setback points on all profiles were joined together on an aerial plan to produce a first approximation of the setback line. A minimum distance of 40 ft from the present dynamic beachface was adopted where the setback line was less stringent. This was done to provide for dynamic beach fluctuations. Further adjustment was made to accommodate estimated erosion rate. Where a dune system was present, the setback line was set at a minimum distance of 25 ft behind the most seaward dune.

Finally, the setback line was smoothed to avoid discontinuities and referenced to the monumented baseline. The recommended setback line has taken into account dynamic features such as waves, tides, storm surges, and erosion rates, as well as topographic features such as dune elevation, beach slope, and beach material.

(b) Coastal setback line on Kiawah Island, South Carolina. Kiawah Island is a sparsely-populated barrier island near Charleston, South Carolina, U.S.A. Geomorphological interpretation of beach processes was carried out to establish coastal setback lines on Kiawah, prior to planned development. 9/

Winds were found to be more frequent from the southwest, although northeast winds were predominant and stronger. Average wave height was 50 cm, and wave refraction around adjacent tidal inlets was determined to be effective in causing shifting zones of erosion. Spring tidal range was about 1.8 m, and strong tidal currents were measured in the inlets. Longshore sediment transport was predominantly from northeast to southwest. The recurved spit at the southern end of the island was determined to have been growing at a rate of about 40 m per year since 1949. Sand size was everywhere fine to very fine.

Historical evaluation of records of the shoreline showed periodic breaching of the spit as the Kiawah River migrated its course. Areas of accretion and erosion were identified, and it was possible to predict partial erosion of the eastern shoreline. Evidence of this possibility was confirmed by the presence of small trees and bushes in the intertidal zone, the truncation of beach ridges, and marsh clays cropping out on the beachface. Rapid accretion in the northeast was related to the construction of jetties in nearby Charleston Harbour 100 years previously. Recent changes were analysed from a series of

---

9/ M. O. Hayes, Development of Kiawah Island, South Carolina, Coastal and Ocean Division of ASE, Charleston, Coastal Sediments, 1977, pp. 828-847.

aerial photos and indicated that the east end of the island was extremely unstable, while the middle part was essentially stable.

Geomorphological considerations determined four major subdivisions: (i) a beach zone, (ii) three tidal inlets, (iii) beach ridge complexes in the interior, and (iv) a saltmarsh/tidal channel area surrounding the landward portion of the barrier.

Five major morphological types, in order of decreasing stability, were found in the beach zone: (i) the stable western beach of the island, (ii) the prograding beach near the east end, (iii) the recurved spit beach, (iv) the washover, transgressing beach near the east end, and (v) the inlet-influenced beach. Consideration of these types was judged important for determining a setback line.

A one-year survey of beach profiles was conducted to determine the beach processes that were active. Areas of instability were determined. Calculations were made of the volume of sand in two nearby tidal deltas. It was determined that the large sand bodies were active in causing wave refraction which shaped Kiawah Island. Saltmarshes and tidal creeks were found to surround the northern part of the island which was generally stable.

It was concluded that the east end of Kiawah Island is extremely unstable and subject to rapid and drastic erosion and deposition. The middle of the island is relatively stable. Three particular areas were pointed out:

- Northeast end of the island: this was judged capable of erosional and depositional changes in the order of tens of metres per year.

- Kiawah River recurved spit: although the wide recurved spit was accreting at the rate of 40 m per year, it was in danger of being breached at the neck by a storm. Hence, it was judged to be unsafe for development.

- Prograding, reclaimed inlet area: situated at the extreme east end of the island, this was judged to be an area of potential property loss, since storm surges might reopen some of the old inlets.

Setback lines were drawn for the island to satisfy the above findings. They were set at the seaward edge of the mature maritime forest in the stable, central part of the island. Since the time of this study, the several findings have been confirmed. Development has proceeded according to the recommended setback line, and no property damage due to erosion took place during a 1979 hurricane.

(c) Stabilization of Singapore shoreline. This is a case involving the application of crenulate-shaped bays to the stabilization of a reclaimed shoreline on Singapore Island. 10/

Crenulate-shaped bays are ubiquitous on all coastlines. They occur wherever a sedimentary beach is set between two headlands and is exposed to an oblique wave approach. The result of wave action is a bay with a half-heart shape. In its equilibrium situation, the curvature of the bay is defined by a logarithmic spiral. The constant in the descriptive log spiral

---

10/ R. Silvester and S. K. Ho, Use of Crenulate-shaped Bays to Stabilize Coasts, in Proceedings of Coastal Engineering Conference, Vancouver, Vol. II. 1972, pp. 1347-1365.

equation has a specific relationship to the angle which the wave crests make with alignment of the headlands. This relationship can be used as a measure of the equilibrium of any such bay. The limit of erosion, in case sand supply is cut off, can thus be determined. Wave propagation at the shoreline of equilibrium crenulate-shaped bays is normal to the beach at every point.

The stable bay principle had been applied in 1971 to a reclamation project in Singapore, east of the main harbour. To protect the reclaimed land, construction of a rock revetement with seawall capstone had been commenced along the straight shoreline. Rapid erosion followed and the structures were undermined.

The Singapore Housing and Development Board decided to apply the concept of the crenulate-shaped bay to stabilize coastlines. A series of offshore breakwaters or headlands were placed in front of the reclaimed shoreline to prevent erosion. In between the protected areas, bays were progressively formed. Headlands were constructed near the low-water level in the form of gabions (stones in steel mesh baskets) and rip-rap (compacted soil mounds faced with layers of rock).

The gabions had a capacity of $2m^3$ and were filled with stones of about 30 cm in dimension. Four or more layers were used and emerged 30 cm above high-water line. They extended 30 m along the shoreline and were located 60 m from the proposed straight shoreline.

The rip-rap was constructed in the dry and given a seaward slope of 1:2. Layers of stones of increasing size were placed on top, to a height of about 1 m above high-water line. The crown of the rip-rap was about 35 m long and about 70 m from the straight shoreline.

The development of crenulate-shaped bays as a method of beach stabilization was judged economical; in addition, it provided sandy shorelines for the housing developments nearby and calm embayments where children could swim in safety.

The usefulness of the application of setback lines to artificially created crenulate-shaped bays lies in the fact that eroding shorelines can be modified to equilibrium shapes. The indentation to be expected in a straight beach is a function of the spacing of the headlands. Thus, setback lines can be confidently located behind the new proposed beach.

(d) Charente Maritime, France: coastal setback lines for coastal zone management. Coastal construction setback lines are useful for planning multiple activities to provide the best distribution of the various activities along the coast. The study done for Charente Maritime in France is such an example. 11/

A comprehensive study of 344 km of Atlantic coastline in Charente Maritime was done to determine possible future utilization and areal distribution of activities in the coastal zone. Examination of the morphology, hydrology, and sedimentology of the area was carried out. Maps of the area under study were

---

11/ F. Manaud, Evaluation sommaire des aptitudes des sites à un développement futur, CNEXO-COB, France, 1977.

prepared to represent separtely a number of characteristics, including:

- bathymetry of the offshore area
- degree of stability of the shoreline, showing zones of erosion and accretion
- wave regime, and convergence of wave energy
- current regime
- movement of water masses
- tidal flows
- flushing and residence time of water
- dispersion of pollutants
- turbidity of the water
- biological productivity.

Planned activities were identified, and their environmental requirements were tabulated. Activities studied included:

- fishing
- mariculture
- navigation and ports
- sewage outfalls
- dredging of ports and inlets
- excavation of sand and gravel
- marine parks
- biological reserves.

The tabulated requirements of each activity were examined in the light of the mapped characteristics of the region, and maps were prepared to show the most suitable locations for each of the activities.

After the consideration for the attributes of the various sites, the existing content, and the necessary environmental and socio-economic factors, the different activities occurring in their assigned zones were brought together on a single map which shows the spatial distribution of the activities. The spatial distribution of the activities thus constitutes the setback lines.

5. Conclusion

Once a coastal construction setback line has been adopted, it should be enforced. For this purpose, the line must be clearly defined. A series of surveyor's concrete monuments (each with a brass identification cap on top) should be set up along the beach. These serve as a monumented baseline which can be referenced to the national system of plane co-ordinates. The coastal setback line is then referenced to the monumented baseline and filed with the competent authorities. Aerial photoplans showing the location of each monument should also be placed on file.

The types of 'construction activity' that are not permitted in front of the setback line also need to be defined. They should include for example, erection of any type of structure; excavation; removal of beach material; any attempt to alter ground elevations; any damage to sand dunes or to vegetation growing thereon; driving any vehicle on, over, or across sand dunes; any alteration of structures already existing within the restricted zone.

Any coastal construction activity which is in violation of the above provision may be declared illegal. Such construction should be removed, or excavation filled, within a stated time period by the violator at his cost. Otherwise, the competent authorities should have the right to remove the construction or fill the excavation at the expense of the violator.

Certain types of construction activity need to be exempted from the provisions of the setback line. Exceptions to the rule would include such government and community needs as the construction of pipelines that cross the beach; building of marinas, fishing piers, and harbours; and laying of submarine telephone cables and of stormwater drain pipes.

Any structure that was erected in the restricted zone before the adoption of the coastal setback line would also be granted a variance. Such a structure would be allowed for the remainder of its natural life. However, no alterations of the structure should be permitted.

The authorities would have the right to grant other variances in cases where it can be clearly established by engineering data that erosion would not occur as a result of the construction activity.

Due to the changing nature of the coastline, the setback line should be periodically reviewed to determine the need for modification. Reviews can be made every five or ten years, depending on the degree of instability of the beaches involved.